Culture of Plant Cells, Tissues and Organs

Culture of Plant Cells, Tissues and Organs

Bhupendra Kumar

RANDOM

RANDOM PUBLICATIONS

NEW DELHI (INDIA)

Culture of Plant Cells, Tissues and Organs

ISBN 978-93-5111-218-1

Published in 2014 in India by

RANDOM PUBLICATIONS

4376-A/4B, Gali Murari Lal, Ansari Road
New Delhi-110 002
Phone : +91-11-43580356, +91-11-23289044
e-mail: randomexports@gmail.com, sales@randompublications.com, info@randompublications.com
Reprinted 2018

Type Setting by : Keystoneprintads, Delhi-110051
Printed at : Replika Press Pvt. Ltd.

Preface

Plant Cell Tissue and Organ Culture is an international journal publishing original results of fundamental studies on plant cells tissues and organs in vitro. Plant Tissue Culture refers to the technique of growing plant cells, tissues, organs, seeds or other plant parts in a sterile environment on a nutrient medium.

In plant tissue culture, plants or explants such as pieces of leave, stem or root is cultured in a specific plant medium, which contains essential plant nutrients and hormones. Other plant growth factors like light and temperature are maintained and regulated by using artificial conditions. All the procedures of plant tissue culture are conducted under sterile (aseptic) conditions. The explants then develop stem, roots and leaves. The generated plantlets are hardened before planting in outdoor conditions.

Plant cells are generally larger than animal cells. While animal cells come in various sizes and tend to have irregular shapes, plant cells are more similar in size and are typically rectangular or cube shaped. A plant cell also contains structures not found in an animal cell. Some of these include a cell wall, a large vacuole, and plastids. Plastids, such as chloroplasts, assist in storing and harvesting needed substances for the plant. Animal cells also contain structures such as centrioles, lysosomes, and cilia and flagella that are not typically found in plant cells.

Tissue culture is the process whereby small pieces of living tissue (explants) are isolated from an organism and grown aseptically for indefinite periods on a nutrient medium. For good success in plant tissue culture, it is good to start with explant that is meristems because these cells are capable of rapid proliferation. The used explants include buds, root tips, nodal segments or germinating seeds and these are placed on suitable culture media where they grow into an undifferentiated mass known as Callus. It deals with the culture of the isolated organs (roots) under laboratory conditions (in vitro).

Different names are given depending upon the organ used for the culture. For instance the culture of roots, endosperm, ovary, and ovule are called as root culture, endosperm culture, ovary culture, etc. It was Skoog (1944), who for the first time suggested that the organogenesis could be chemically controlled. Skoog and Miller (1957) also demonstrated that a high ratio of auxin: cytokinin stimulated the formation of root in tobacco callus, but a low ratio of the same induced shoot information.

This book is aimed at all those interested and involved in the results of plants in the environment and is an essential basic reference book for students, academicians and specialists.

I thank all members of my team who have helped in the preparation of the book. My special thanks go to "Random Publications" who have published the book.

– Bhupendra Kumar

Contents

1

Introduction

The cultured cells and tissue can take several pathways to produce a complete plant. Among these, the pathways that lead to the production of true-to-type plants in large numbers are the popular and preferred ones for commercial multiplication. The following terms have been used to describe various pathways of cells and tissue in culture.

REGENERATION AND ORGANOGENESIS

In this pathway, groups of cells of the apical meristem in the shoot apex, axillary buds, root tips, and floral buds are stimulated to differentiate and grow into shoots and ultimately into complete plants. In many cases, the axillary buds formed in the culture undergo repetitive proliferation, and produce large number of tiny plants. The plants are then separated from each other and rooted either in the next stages of micropropagation or in vivo (in trays, small pots or beds in glasshouse or plastic tunnel under relatively high humidity). The explants cultured on relatively high amounts of auxin (*e.g.* (2,4-D, 2,4-dichlorophenoxyacetic acid) form an unorganized mass of cells, called callus. The callus can be further sub-cultured and multiplied.

The callus shaken in a liquid medium produces cell suspension, which can be subcultured and multiplied into more liquid cultures. The cell suspensions form cell clumps, which eventually form calli and give rise to plants through organogenesis or somatic embryogenesis. In some cases, explants *e.g.* leaf-discs and epidermal tissue can also generate plants by direct organogenesis and somatic embryogenesis without intervening callus formation, *e.g.* in orchardgrass. Dactylis glomerata L.. In organogenesis the cultured plant cells and cell clumps (callus) and mature differentiated cells (microspores, ovules) and tissues (leaf discs, inter-nodal segments) are induced to differentiate into complete plants to form shoot buds and eventually shoots, and rooted to form complete plants.

SOMATIC EMBRYOGENESIS

In this pathway, cells or callus cultures on solid media or in suspension cultures form embryo-like structures called somatic embryos, which on germination produce complete plants. The primary somatic embryos are also capable of producing more embryos through secondary somatic embryogenesis.

Although, somatic embryogenesis has been demonstrated in a very large number of plants and trees, the use of somatic embryos in large-scale commercial production has been restricted to only a few plants, such as carrot, date palm, and a few forest trees. Somatic embryos are produced as adventitious structures directly on explants of zygotic embryos, from callus and suspension cultures. Somatic embryos and synthetic seeds (embryos encapsulated in artificial endosperm) hold potential for large-scale clonal propagation of superior genotypes of heterogeneous plants. They have also been used in commercial plant production and for the multiplication of parental genotypes in large-scale hybrid seed production.

In many species, somatic embryos are morphologically similar to the zygotic embryos, although some biochemical, physiological and anatomical differences have been documented. The synthetic auxin, 2,4-D is commonly used for embryo induction. In many angiosperms, *e.g.*, carrot and alfalfa, subculture of cells from 2,4-D containing medium to auxin-free medium is sufficient to induce somatic embryogenesis.

The process can be enhanced with the application of osmotic stress, manipulation of medium nutrients, and reducing humidity. Selection of embryogenic cell lines has also been successfully used. For example, selection for unique morphotypes in grapevine cultures allows production of high quality embryos with predicable frequency. A major problem in large-scale production of somatic embryos is culture synchronization. This is achieved through selecting cells or pre-embryonic cell clusters of certain size, and manipulation of light and temperature, temporary starvation or by adding cell cycle synchronizing chemicals to the medium. Cytokinins seem to play a key role in cell cycle synchronization and embryo induction, proliferation and differentiation. Abscisic acid is crucial in all the stages of somatic development, maturation and hardening.

SYNTHETIC SEEDS

The concept of production and utilization of synthetic seeds (somatic embryo as substitutes for true seeds) was first suggested by Murashige in 1977. Synthetic seeds can be produced either as coated or non-coated, desiccated somatic embryos or as embryos encapsulated in hydrated gel (usually calcium alginate).

Successful utilization of synthetic seeds as propagules of choice requires an efficient and reproducible production system and a high percentage of post-planting conversion into vigourous plants. Artificial coats and gel capsules containing nutrients, pesticides and beneficial organisms have long been thought as substitutes for seed coat and endosperm. However, this technology is still in the developmental stage, and currently cannot compete with the other methods of commercial plant propagation.

Process of Micropropagation

The process of plant micro-propagation aims to produce clones (true copies of a plant in large numbers).

The process is usually divided into the following stages:

Stage 0 – Pre-propagation step or selection and pre-treatment of suitable plants.

Stage I – Initiation of explants-surface sterilization, establishment of mother explants.

Stage II – subculture for multiplication/proliferation of explants.

Stage III – shooting and rooting of the explants.

Stage IV – Weaning/hardening.

These stages are universally applicable in large-scale multiplication of plants. The individual plant species, varieties and clones require specific modification of the growth media, weaning and hardening conditions.

A rule of the thumb is to propagate plants under conditions as natural or similar to those in which the plants will be ultimately grown ex-vitro. For example, if a chrysanthemum variety is to be grown under long day-length for flower production, it is better to multiply the material under long-day length at stages III and IV. There is a wide option to undertake production of plant material up to a limited number of stages. For example, many commercial tissue culture companies undertake production up to Stage III, and leave the remaining stages to others.

Pre-Propagation Stage

The pre-propagation stage (also called stage 0) requires proper maintenance of the mother plants in the greenhouse under disease- and insect-free conditions with minimal dust. Clean enclosed areas, glasshouses, plastic tunnels, and net-covered tunnels, provide high quality explant source plants with minimal infection.

Collection of plant material for clonal propagation should be done after appropriate pretreatment of the mother plants with fungicides and pesticides to minimize contamination in the in vitro cultures. This improves growth and multiplication rates of in vitro cultures. The control of contamination begins with the pretreatment of the donor plants.

They may be prescreened for diseases, isolated and treated to reduce contamination. The explants are then brought to the production facility, surface sterilized and introduced into culture. They may at this stage be treated with antibiotics and fungicides as well as anti-microbial formulations, such as PPM. The explants are then culture indexed for contamination by standard microbiological techniques, which are occasionally supplemented with tests based on molecular biology or other techniques.

Stage I

This stage refers to the inoculation of the explants on sterile medium to initiate aseptic culture. Initiation of explants is the very first step in micropropagation. A good clean explant, once established in an aseptic condition, can be multiplied several times; hence, explant initiation in an aseptic condition should be regarded as a critical step in micropropagation. More than often, explants fail to establish and grow, not due to the lack of a suitable medium but because of contamination. The explants are transferred to in vitro environment, free from microbial contaminants. The process requires excision of tiny plant pieces and their surface sterilization with chemicals such as sodium hypochlorite, ethyl alcohol and repeated washing with sterile distilled water before and after treatment with chemicals. After a short period of culture, usually 3 to 5 days, the contaminated explants are discarded.

The surviving explants showing growth are maintained and used for further subculture. In herbaceous plants *e.g.* potato, chrysanthemum, carnation, streptocarpus, strawberry, and African violet; the explant sources are meristems, apical- and axillary buds, young seedlings, developing young leaves and petioles, and unopened floral buds. The following low cost options can be adapted to initiate explants:

Sterile Instrument Technique

This method assumes that most of the deep-seated meristems and those covered by leaves or other integuments (*e.g.* floral bracts) are sterile. In this procedure, the explant is washed with sterile water, rinsed in ethanol, and instruments are sterilized every time they touch the surface of the explant, and the explant is moved to a new location on the dissection stage.

Surface Sterilization Technique

This is by far the most commonly used method. The explants are washed in sterile water, rinsed in ethanol, and surface sterilization is achieved by using chemicals with chlorine base. Calcium or sodium hypochlorite based solutions, 1–3 per cent (v/v) are usually used for soft herbaceous materials. A cheap and ready-made sterilant is 5-7 per cent solution of 'Domestos'- a toilet disinfectant which contains 10.5 per cent v/v sodium hypochlorite, 0.3 per cent sodium carbonate, 10.0 per cent sodium chloride and 0.5 per cent (w/v) sodium hydroxide and a patented thickener). The explants are washed in sterile distilled water before and after sterilization. Other surface sterilants used include mercuric chloride (avoid its use as far as possible, since it is highly toxic), hydrogen peroxide, and potassium permanganate.

For Soft Tissues

- Wash explants from perennial plants for 1-2 hr in tap water. Eliminate this step for material from glasshouse grown plants.
- Wash in sterile distilled water three to four times for 5 to 10 minutes each.
- Dip in 95 per cent ethanol for 3 to 5 seconds.
- Wash once again with sterile distilled water for 5 minutes.
- Surface-sterilize in 5 per cent 'Domestos' (v/v) for 20-25 minutes.
- Wash with sterile distilled water three times for 10 minutes each.
- Drain water droplets by placing on pre-sterilized blotting paper.
- Transfer explants singly to the medium.

N.B.: Sterilize forceps each time to transfer explants to avoid cross-contamination.

For woody stems (e.g. roses, hardy shrubs, and trees):

- Collect stems, shoots, buds and store at 5 oC till needed.
- Rinse in ethanol for 3 to 5 seconds.
- Rinse in 1- per cent sodium hypochlorite (20 per cent bleach) for 10 minutes.
- Place lower parts of stems in flasks in 2 per cent sucrose and 200 PPM 8-hydroxyquinoline citrate at 23+2 oC. For items collected in September/October, add 50-PPM GA3.

After that l0 PPM GA will help break the dormancy:

- Re-cut the bottom of stem and replace the solution after 2 days.
- Excise the softwood from the developed shoots and use material for explants or for rooting.
- Surface-sterilize as in the above protocol.

Do not forget to sterilize forceps and scalpel every time for the transfer of explants to fresh solutions. Use sterile containers in the protocol of surface sterilization. If explants become brown or pale at the end of the protocol, reduce the strength of 'Domestos' to 2.5 per cent. Alternatively, dip explants in 10 per cent 'Domestos' for 2 minutes and then proceed to surface sterilize with 3-5 per cent 'Domestos' for 20 minutes. If basal contamination is observed after 2-3 days of culture, explants can sometimes be rescued by removing the basal end by making a single cut with a sharp scalpel and re-culturing on fresh medium.

Stage II

Stage II is the propagation phase in which the explants are cultured on the appropriate media for multiplication of shoots. The primary goal is to achieve propagation without losing the genetic stability. Repeated culture of axillary and adventitious shoots, cutting with nodes, somatic embryos and other organs from Stage I leads to multiplication of propagules in large numbers. The propagules produced at this stage can be further used for multiplication by their repeated culture. Sometimes it is necessary to subculture the in vitro derived shoots onto different media for elongation.

Stage III

The in vitro shoots obtained at Stage II are rooted to produce complete plants. If the proliferated material consists of bud-like structures (*e.g.* orchids) or clumps of shoots (banana, pineapple), they should be separated after rooting and not before. Many plants (*e.g.* banana, pineapple, roses, potato, chrysanthemum, strawberry, mint, several grasses and many more) can be rooted on half-strength-MS medium without any growth-regulators. Good sturdy well-rooted plants are essential for high survival during weaning and later transfer to soil. This stage is labour intensive and expensive. The process of in vitro rooting has been estimated to account for approximately 35–75 per cent of the total cost of production. Efforts should be made to combine rooting and acclimatization stages.

Stage IV

At this stage, the in vitro micropropagated plants are weaned and hardened. This is the final stage of the tissue culture operation after which the micropropagated plantlets are ready for transfer to the greenhouse. Steps are taken to grow individual plantlets capable of carrying out photosynthesis. The hardening of the tissue-cultured plantlets is done gradually from high to low humidity and from low light intensity to high intensity conditions. If grown on solid medium, most of the agar can be removed gently by rinsing with water. Plants can be left in shade for 3 to 6 days where diffused natural light conditions them to the new environment. The plants are then transferred to an appropriate substrate (sand, peat, compost, etc.), and gradually hardened. Low-cost options include the use of plastic domes or tunnels, which reduces the natural light intensity and maintains high relative humidity during the hardening process. If the plants are still joined together after rooting, these should be planted as bunches in the soil and separated after 6 to 8 weeks of growth.

METHODS AND TECHNIQUES OF PLANT TISSUE CULTURE

In plant tissue culture, plants or explants such as pieces of leave, stem or root is cultured in a specific plant medium, which contains essential plant nutrients and hormones. Other plant growth factors like light and temperature are maintained and regulated by using artificial conditions. All the procedures of plant tissue culture are conducted under sterile (aseptic) conditions. The explants then develop stem, roots and leaves. The generated plantlets are hardened before planting in outdoor conditions.

To start with plant tissue culture, one should wipe his/her hands, forceps and other equipments with alcohol or other sterilizer to prevent microbial contamination. Following this, the explants are surface sterilized by using chemical solutions such as bleach or alcohol. Though mercury chloride is an effective sterilizer, it is rarely used due to its potential toxicity. After sterilization, the explants are introduced into a plant medium, which can be either solid or liquid. The cells divide and differentiate into plant parts, thus giving rise to a complete plant. Speaking about the plant medium, liquid type is prepared by mixing inorganic salts, organic nutrients (sucrose), vitamins, minerals and plant growth regulators (auxin and cytokinin). In addition to these ingredients, solid medium contains gelling agent (agar). Nutrient and plant hormones amount vary, depending on the objective of plant tissue culture. For example, in order to induce more roots, auxin amount should be high. Plant tissue culture techniques include the culture of protoplast (a cell without cell wall), meristem, node, anther, ovule, embryo and seed.

APPLICATIONS OF PLANT TISSUE CULTURE

Plant tissue culture has wide applications in agriculture. One major advantage of plant tissue culture is the production of disease and/or pest resistant varieties, thus indirectly increasing the crop yield. Commercially, it is used directly for the propagation of plants that are hard to propagate in natural conditions. Horticultural plants such as orchids, roses, banana, strawberries, potatoes, apples, etc. are successfully cultured in in-vitro conditions. Plants with valuable secondary products are grown in the controlled conditions by using plant tissue culture technique. It is also used for the propagation of medicinal herbs on a large scale. With plant tissue culture, it is possible to generate virus-free plantlet of vegetatively propagated plants.

In experimental biology such as plant breeding, cell biology, biotechnology and genetics, plant tissue culture is applied in order to solve plant related problems. It allows screening of cells for the desirable characters such as early fruit bearing, disease resistance and drought resistance. Another important application is generation of a novel hybrid by crossing two distantly related species having advantageous traits. In-vitro fertilization and/or pollination of plants is possible, irrespective of the hindrances in natural conditions. Overall, plant tissue culture is used for the conservation of germplasm.

PLANT TISSUE CULTURE MEDIA

Plant Tissue Culture refers to the technique of growing plant cells, tissues, organs, seeds or other plant parts in a sterile environment on a nutrient medium.

Culture media used for in vitro cultivation of plant cells are composed of following basic components:

- *Complex mixture of salts*: Essential elements, or mineral ions.

- *Organic supplement*: Vitamins and/or amino acids.
- *Carbon source*: Usually sugar sucrose.
- Gelling agents
- Plant Growth Regulators
- Antibiotics

Complex Mixture of Salts

These include essential elements or mineral ions important for plant nutrition and their physiological function.

The essential elements can further be divided into the following categories:

- Macroelements (or macronutrients)
- Microelements (or micronutrients)
- Iron source

Macroelements

Elements are required in large amounts for plant growth and development. Nitrogen, phosphorus, potassium, magnesium, calcium and sulphur (and carbon, which is added separately) are regarded as macroelements. These elements comprise at least 0.1% of the dry weight of plants.

Microelements

Elements are required in trace amounts for plant growth and development. Manganese, iodine, copper, cobalt, boron, molybdenum, iron and zinc are regarded as microelements, although other elements like aluminium and nickel are frequently found in some formulations.

Iron Source

Iron is usually added in the medium as iron sulphate, although iron citrate can also be used. Ethylenediaminetetraacetic acid (EDTA) is usually used in conjunction with the iron sulphate. The EDTA complexes with the iron so as to allow the slow and continuous release of iron into the medium. Uncomplexed iron can precipitate out of the medium as ferric oxide.

Organic Supplements

These include vitamins and amino acids. Two vitamins, *i.e.*, thiamine (vitamin B1) and myoinositol (a vitamin B) are essential for the culture of plant cells in vitro. However, other vitamins are often added to for historical reasons. The most commonly used amino acid is glycine.

However, arginine, asparagine, aspartic acid, alanine, glutamic acid, glutamine and proline are also used. Amino acids provide a source of reduced nitrogen and, like ammonium ions, uptake causes acidification of the medium. Casein hydrolysate can be used as a source of a mixture of amino acids.

Carbon Source

The most commonly used carbon source is sucrose. It is readily assimilated and relatively stable. Other carbohydrates like glucose, maltose, galactose and sorbitol can also be used and may prove better than sucrose in specialized circumstances.

Gelling Agents

Plant tissue culture media can be used in either liquid or 'solid' forms, depending on the type of culture being grown. Agar, produced from seaweed, is the most common type of gelling agent, and is ideal for routine applications. For more demanding applications, a range of purer gelling agents are available. Purified agar or agarose can be used, as can a variety of gellan gums.

Plant growth regulators: Specific media manipulations can be used to direct the development of plant cells in culture due to plasticity and totipotency. Plant growth regulators are the critical media components in determining the developmental pathway of the plant cells.

There are five main classes of plant growth regulator used in plant cell culture, namely:

1. Auxins
2. Cytokinins
3. Gibberellins
4. Abscisic acid
5. Ethylene

Auxins

Auxins promote both cell division and cell growth. IAA (indole-3-acetic acid) is the most important naturally occurring auxin but its use in plant tissue culture media is limited because it is unstable to both heat and light. 2,4-Dichlorophenoxyacetic acid (2,4-D) is the most commonly used auxin and is extremely effective in most circumstances.

Cytokinins

Cytokinins promote cell division. Of the naturally occurring cytokinins, only zeatin and 2iP (2-isopentyl adenine have some use in plant tissue culture media. The synthetic analogues, kinetin and BAP (benzylaminopurine), are used more frequently. Non-purine-based chemicals, such as substituted phenylureas, are also used as cytokinins in plant tissue culture media.

Gibberellins

Gibberellins are involved in regulating cell elongation, in determining plant height and fruit-set. Only a few of the gibberellins like GA3 are used in plant tissue culture media.

Abscisic Acid

It is used in plant tissue culture to promote distinct developmental pathways such as somatic embryogenesis. Abscisic acid (ABA) inhibits cell division.

Ethylene

Ethylene is associated with controlling fruit ripening in climacteric fruits, and its use in plant tissue culture is not widespread. Some plant cell cultures produce ethylene, which, if it builds up sufficiently, can inhibit the growth and development of the culture.

Antibiotics

Antibiotics are substances produced by certain microorganisms that suppress the growth of other microorganisms and eventually destroy them.

Their applications include:

- Suppresses bacterial infections in plant cell and tissue culture.
- Suppresses mould and yeast infections in cell cultures.
- Eliminates Agrobacterium species after the transformation of plant tissue.

These antibiotics can be divided into different classes on the basis of chemical structure and their mechanism of action:

- Inhibitors of Bacterial Cell Wall Synthesis.
  e.g. â-lactam antibiotics, Penicillins and Cephalosporins.
- Antibiotics that affect Cell Membrane permeability.
  - Antibacterial *e.g.* Colistin Sulphate, Polymixin B Sulphate, Gramicidin
  - Antifungal *e.g.* Amphotericin B, Nystatin, Pimaricin
- Bacteriostatic Inhibitors of Protein Antibiotics that affect the function of 30 S or 50 S ribosomal subunits to cause a reversible inhibition of protein synthesis. *e.g.* Chloramphenicol, Chlortetracycline HCl, Clindamycin HCl, Doxycycline HCl, Erythromycin, Lincomycin HCl, Oxytetracycline HCl, Spectinomycin sulphate, Tetracycline HCl, Tylosin tartrate, Lincomycin HCl
- Bactericide Inhibitors of Protein Synthesis Antibiotics that bind to the 30 S ribosomal subunit and alter protein synthesis which eventually leads to cell death. This group includes:
  - *Aminoglycosides*: Apramycin, Butirosine, Gentamicin, Kanamycin, Neomycin, Streptomycin, Tobramycin.
  - *Inhibitors of Nucleic Acid Metabolism*: *e.g.* Rifampicin, Mitomycin C and Nalidixic acid.
  - Antimetabolites: Antibiotics, which block specific metabolic steps that are essential to microorganisms *e.g.* Metronidazole, Miconazole, Nitrofurantoin, Trimethoprim and Sulpho-methoxazole.
  - Nucleic Acid Analogs, which inhibit enzymes essential for DNA synthesis. *e.g.* 5-Fluorouracil, Mercaptopurine.

2

Plant Growth and Development

THE ANALYSIS OF PLANT GROWTH

New cells form continually in the apical meristems. Cells enlarge slowly in the apical meristem and more rapidly in the subapical regions. The resulting increase in cell volume can range from severalfold to one hundredfold, depending on the species and environmental conditions. Classically, plant growth has been analyzed in terms of cell number or overall size (*i.e.*, mass). However, these measures tell only part of the story. Tissue growth is neither uniform nor random. The derivatives of the apical meristems expand in predictable and site-specific ways, and the expansion patterns in these subapical regions largely determine the size and shape of the primary plant body. The total growth of the plant can be thought of as the sum of the local patterns of cell expansion.

The analysis of the motions of cells or "tissue elements" (and the related problem of cell expansion) is called *kinematics*. In this section we will discuss both the classical definitions of growth and the more modern, kinematic approach. As we will see, the advantage of the kinematic approach is that it allows one to describe the growth patterns of organs mathematically in terms of the expansion patterns of their component cells.

Plant Growth Can Be Measured in Different Ways: Growth in plants is defined as an irreversible increase in volume. The largest component of plant growth is cell expansion driven by turgor pressure. During this process, cells increase in volume manyfold and become highly vacuolate. However, size is only one criterion that may be used to measure growth.

Growth also can be measured in terms of change in fresh weight—that is, the weight of the living tissue—over a particular period of time. However, the fresh weight of plants growing in soil fluctuates in response to changes in water status, so this criterion may be a poor indicator of actual growth. In these situations, measurements of dry weight are often more appropriate. Cell number is a common and convenient parameter by which to measure the growth of unicellular organisms, such as the green alga *Chlamydomonas*. In multicellular plants, however, cell number can be a misleading growth measurement because cells can divide without increasing in volume.

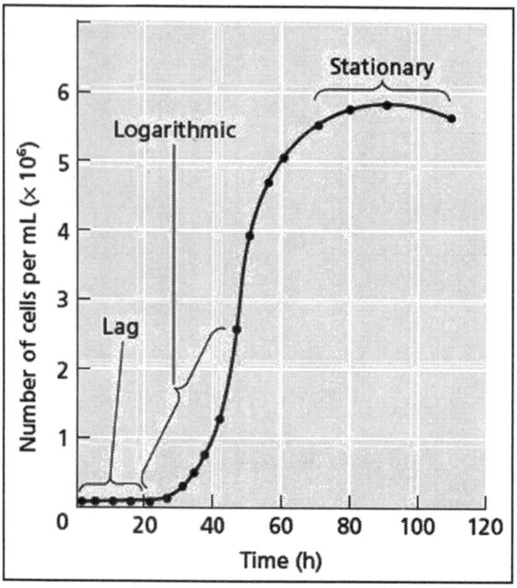

Fig. Growth of the unicellular green alga *Chlamydomonas*.

Growth is assessed by a count of the number of cells per milliliter at increasing times after the cells are placed in fresh growth medium. Temperature, light, and nutrients provided are optimal for growth. An initial lag period during which cells may synthesize enzymes required for rapid growth is followed by a period in which cell number increases exponentially. This period of rapid growth is followed by a period of slowing growth in which the cell number increases linearly. Then comes the stationary phase, in which the cell number remains constant or even declines as nutrients are exhausted from the medium.

For example, during the early stages of embryogenesis, the zygote subdivides into progressively smaller cells with no net increase in the size of the embryo. Only after it reaches the eight-cell stage does the increase in volume begin to mirror the increase in cell number. Because the zygote is an especially large cell, this lack of correspondence between an increase in cell number and growth may be unusual, but it points out the potential problem in equating an increase in cell number with growth. Although cell number may not always be a reliable measure of plant growth, under most circumstances dividing cells, particularly in meristems, double in volume during their cell cycle. Therefore, an increase in cell number, such as the increase brought about by the activity of the apical meristems, does contribute to plant growth. However, the largest component of plant growth is the rapid cell expansion that occurs in the subapical region after cell division ceases.

Because all the cells of the plant axis elongate under normal conditions, the greater the number of cells produced by the apical meristem, the longer the axis will be. For example, when Arabidopsis plants are transformed with a gene that encodes cyclin, a key component of the cell cycle regulatory machinery, the cells of the apical meristem progress through their cell cycles more rapidly, so more cells form per unit time. As a result, the roots of these transgenic plants have more cells and are substantially longer than the roots of wild-type plants grown under similar conditions. New cells form continually in the apical meristems. With each new round of cell division and associated cell expansion, the older derivatives are displaced a small distance from the apex. As the cells recede farther from the apex, the rate of displacement is greatly accelerated. By viewing plant growth as a process of cell displacement from the apex, we can apply the principles of kinematics.

THE PRODUCTION OF CELLS

Moving fluids such as waterfalls, fountains, and the wakes of boats can generate specific forms. The study of the motion of fluid particles and the shape changes that the fluids undergo is called kinematics. The ideas and numerical methods used to study these fluid forms are useful for characterizing meristematic growth. In both cases, an unchanging form is produced even though it is composed of moving and changing elements. An example of an unchanging form composed of changing and displaced elements in plants is the hypocotyl hook of a dicot, such as the common bean. As the bean seedling emerges from the seed coat, the apical end of the hypocotyl bends back on itself to form a hook. The hook is thought to protect the seedling apex from damage during growth through the soil. During seedling growth (in soil or dim light) the hook migrates up the stem, from the hypocotyl into the epicotyl and then to the first and second internodes, but the form of the hook remains constant.

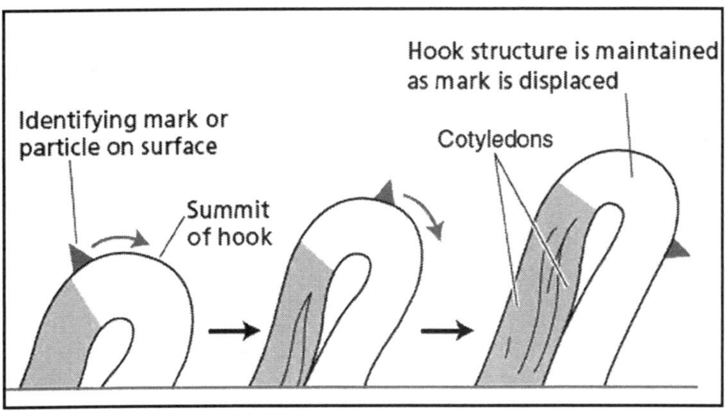

Fig. The dicot Hypocotyl hook is an example of a Constant form Composed of changing Elements.

The hooked form is maintained over time, while different tissues first curve and then straighten as they are displaced from the seedling apex during growth. If a mark is placed at a fixed point on the surface, it will be displaced (indicated by the arrow), appearing to flow through the hook over time. If we mark a specific epidermal cell on the seedling stem located close to the seedling apex, we can watch it as it flows into the hook summit, then down into the straight region below the hook.

The mark is not crawling over the plant surface, of course; plant cells are cemented together and do not experience much relative motion during development. The change in position of the mark relative to the hook implies that the hook is composed of a procession of tissue elements, each of which first curves and then straightens as it is displaced from the plant apex during growth. The steady form is produced by a parade of changing cells. A root tip is another example of a steady form composed of changing tissue elements. Here, too, the form is observed to be steady only when distance is measured from the root tip. A region of cell division occupies perhaps 2 mm of the root tip. The elongation zone extends for about 10 mm behind the root tip. Phloem differentiation is first observed beginning at 3 mm from the tip, and functional xylem elements may be seen at about 12 mm from the tip.

A marked cell near the tip will seem to flow first through the region of cell division, then through the elongation zone and into the region of xylem differentiation, and so on. This shifting implies that developing tissue elements first divide and elongate, and then differentiate. In an analogous fashion, the shoot bears a succession of leaves of different developmental stages. During a period of 24 hours, a leaf may grow to the same size, shape, and biochemical composition that its neighbour had a day earlier. Thus, shoot form is also produced by a parade of changing elements that can be analyzed with kinematics. Such an analysis is not merely descriptive; it permits calculations of the growth and biosynthetic rates of individual tissue elements (cells) within a dynamic structure.

Tissue Elements Are Displaced during Expansion

As we have seen, growth in shoots and roots is localized in regions at the tips of these organs. Regions with expanding tissue are called growth zones. With time, meristems move away from the plant base by the growth of the cells in the growth zone.

If successive marks are placed on the stem or root, the distance between the marks will change, depending on where they are within the growth zone. In addition, all of these marks will move away from the tip of the root or shoot, but their rate of movement will differ depending on their distance from the tip. From another perspective, if you were to stand at the tip of a root that had marks placed at intervals along the axis, you would see that all marks would move farther away from you with time. The reason is that discrete regions on the plant axis experience displacement as well as expansion during growth and development.

As Regions Move Away from the Apex, Their Growth Rate Increases

As a given region of the plant axis moves away from the apex, its growth velocity increases (the rate of elongation accelerates) until a constant limiting velocity is reached equal to the overall organ extension rate. The reason for this increase in growth velocity is that with time, progressively more tissue is located between the moving particle and the apex, and progressively more cells are expanding, so the particle is displaced more and more rapidly. In a rapidly growing maize root, a tissue element takes about 8 hours to move from 2 mm (the end of the meristematic zone) to 12 mm (the end of the elongation zone).

Beyond the growth zone, elements do not separate; neighboring elements have the same velocity (expressed as the change in distance from the tip per unit of time), and the rate at which particles are displaced from the tip is the same as the rate at which the tip moves through the soil. The root tip of maize is pushed through the soil at 3 mm h^{-1}. This is also the rate at which the nongrowing region recedes from the apex, and it is equal to the final slope of the growth trajectory.

The Growth Velocity Profile Is a Spatial Description of Growth

The velocities of different tissue elements are plotted against their distance from the apex to give the spatial pattern of growth velocity, or growth velocity profile. Growth velocity increases with position in the growth zone. A constant value is obtained at the base of the growth zone. The final growth velocity is the final, constant slope of the growth trajectory equal to the elongation rate of the organ, as discussed in the previous section. In the rapidly growing maize root, the growth velocity is 1 mm h^{-1} at 4 mm, and it reaches its final value of nearly 3 mm h^{-1} at 12 mm.

(A) The growth velocity profile plots the velocity of movement away from the tip of points at different distances from the tip. This tells us that growth velocity increases with distance from the tip until it reaches a uniform velocity equal to the rate of elongation of the root. (B) The relative elemental growth rate tells us the rate of expansion of any particular point on the root. It is the most useful measure for the physiologist because it tells us where the most rapidly expanding regions are located.

If the growth velocity is known, the relative elemental growth rate, which represents the fractional change in length per unit time, can be calculated. The relative elemental growth rate shows the location and magnitude of the extension rate and can be used to quantify the effects of environmental variation on the growth pattern.

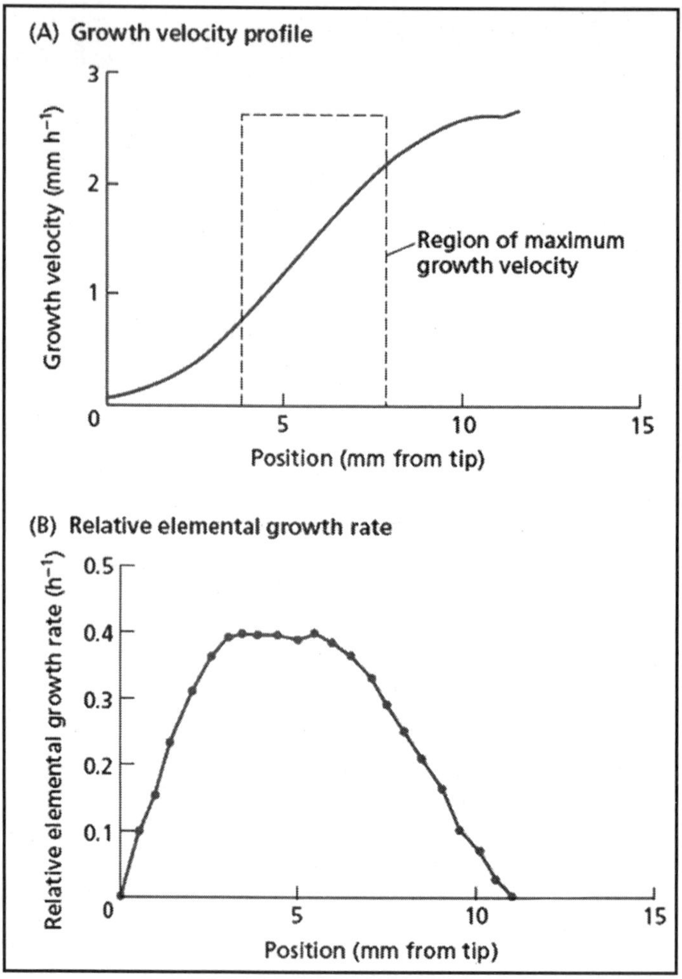

Fig. The Growth of the Primary root of Zea mays (maize) can be represented Kinematically by two related growth curves.

THE RELATIVE ELEMENTAL GROWTH RATE

One way to characterize the growth pattern along an axis mathematically is to plot the distance of a tissue element from the apex versus time. The resulting curve is called the growth trajectory. The slope of the curve at any given point is equivalent to the growth velocity at that region of the axis. With the apex as the reference point, the growth trajectory is concave. Marked tissue elements at different locations may be followed simultaneously to generate a family of growth trajectories. Note that distance is measured from the root tip. This is termed a material specification of growth because real or material particles are followed. As the regions of the plant axis move away from the apex, their growth velocity increases (*i.e.*, the cells accelerate) until a constant limiting velocity is reached equal to the overall organ extension rate. The reason for this increase in growth velocity is that with time, progressively more tissue is located between the moving particle and the apex, and progressively more cells are expanding, so the particle is displaced more and more rapidly.

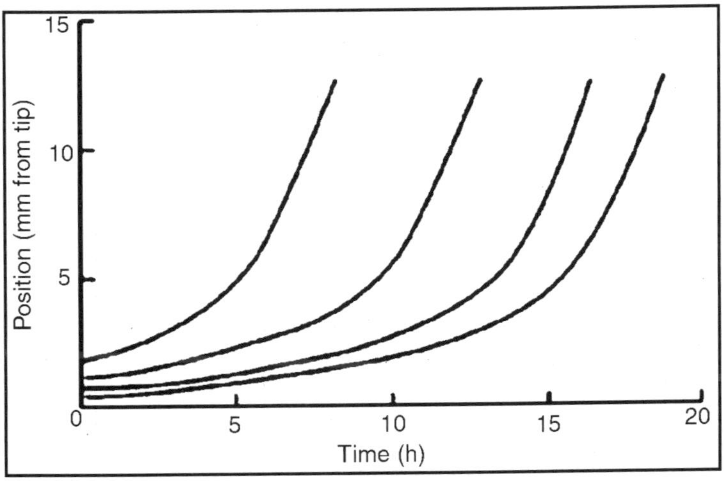

Fig. Growth Trajectories of *Zea mays* (maize):Growth Trajectories plot the positions of points spaced along the root axis as a Function of time.

In a rapidly growing maize root, a tissue element takes about 8 hours to move from 2 mm (the end of the meristem) to 12 mm (the end of the elongation zone). The root tissue element accelerates through this region. Beyond the growth zone, elements do not separate; neighboring elements have the same velocity, and the rate at which particles are displaced from the tip is the same as the rate at which the tip is pushed away from the soil surface.

The root tip of maize is pushed through the soil at 3 mm h^{-1}. This is also the rate at which the nongrowing region recedes from the apex, and it is equal to the final slope of the growth trajectory. The growth trajectory is a material description of growth because an individual element is followed through time. The growth trajectory provides a way to convert spatial information into a developmental time course. For example, if the amount of suberin in the root increases with position, the growth trajectory will give information on the time required for the observed accumulation of suberin in cells moving between any two spatial points.

To characterize the local expansion rate, we must consider the growth velocities at the apical and basal ends of a small segment of tissue. If the apical end of the segment is moving faster than the basal end, the segment is elongating. The velocity difference, divided by the segment length, gives the relative growth rate of the segment. Now, imagine that the segment shrinks to a point x. To find the relative growth rate, r, at point x, we can use calculus:

The value of r is given by the velocity gradient, dv/dx, where v is the growth velocity. This measure of the local growth rate is called the relative elemental growth rate.

It represents the fractional change in length per unit time and has units of h^{-1}. If the growth velocity is known, the relative elemental growth rate can be evaluated by differentiation of the growth velocity with respect to position. Or we can approximate r by measuring the relative growth rates of closely spaced segments of initial length L. For each segment, $r = (1/L)(dL/dt)$.

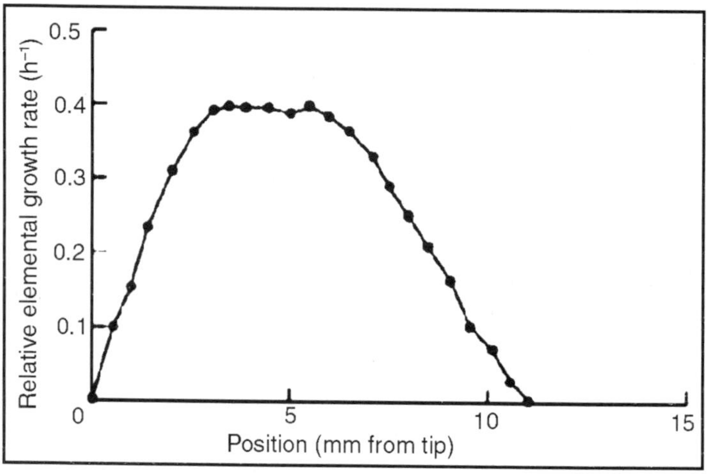

Fig. Relative Elemental Growth rate of *Zea mays* (maize).

The relative elemental growth rate tells us the rate of expansion of any particular point on the root. It is the most useful measure of root growth rate for the physiologist because it tells where the most rapidly expanding regions are located. The relative elemental growth rate profile, $r(x)$, is a spatial description of the expansion pattern in the growth zone. In the maize root, the relative elemental growth rate has a maximum of 0.40 h^{-1} at 4 mm and drops to zero by 12 mm. That is, the tissue element at 4 mm is expanding 40% h^{-1}, while tissue beyond 12 mm has ceased expansion. Again, to understand the profile in physiological terms, we must consider that individual cells expand more and more rapidly as they are displaced to 4 mm and then less and less rapidly as they are displaced through the basal part of the growth zone. The relative elemental growth rate profile shows the location and magnitude of the extension rate and can be used to quantify the effects of environmental variation on the growth pattern. It is also the basis for studying the physiology of growth. The graph of $r(x)$ can be compared to the distribution pattern of any substance thought to be regulating the growth rate. A close correlation between the two curves would provide circumstantial evidence for the role of the substance as a limiting factor for growth.

EMBRYONIC DORMANCY

Embryo Maturation Requires Specific Gene Expression

The Arabidopsis embryo enters dormancy after it has generated about 20,000 cells. Dormancy is brought about by the loss of water and a general shutting down of gene transcription and protein synthesis—not only in the embryo, but also throughout the seed.

To adapt the cell to the special conditions of dormancy, specific gene expression is required. For example, the *A*bscisic *A*cid *I*nsensitive*3* (*ABI3*) and *FUSCA3* genes are necessary for the initiation of dormancy and are sensitive to the hormone abscisic acid, which is the signaling molecule that initiates seed and embryo dormancy.

ABI3 also controls the expression of genes encoding the storage proteins that are deposited in the cotyledons during the maturation phase of embryogenesis.The *LE*AFY *C*OTYLEDON*1* (*LEC1*) gene also is active in late embryogenesis. Because *lec1* mutants cannot survive desiccation and do not enter dormancy, the embryos die unless they are rescued through isolation before desiccation occurs. The rescued embryos will germinate in culture and produce fertile plants, which are like wild-type plants except that they lack the 7S storage protein and they have leaflike cotyledons with trichomes on their upper surface.

The normal appearance and development of the mature *lec1* mutants indicates that the *LEC1* gene is required only during embryogenesis. Although the most obvious defects of the *lec1* mutants are seen only in the maturation phase embryo, mRNA from *LEC1* gene expression can be detected throughout embryogenesis. It has been proposed that *LEC1* is a general repressor of vegetative development and its expression is necessary throughout embryogenesis.

POLARITY OF FUCUS ZYGOTES

Plant cells perceive their position within a tissue and assume the polarity appropriate for that tissue. In addition, cells can perceive and act on information coming from their environment to alter their polarity and redirect growth when necessary. Cellular polarity must be established to orient subsequent expansion. Cells that are initially apolar, such as zygotes, spores, and pollen grains, later become polarized and initiate tip growth and have been especially valuable in the analysis of the induction and establishment of polarity.

A wide variety of external gradients, including temperature, light intensity, pH, ions, and electric-potential gradients, can polarize and orient the axis of growth of these cells. Thus, polarity can be experimentally manipulated without the complexities associated with multicellular tissues.

Much of our understanding of the establishment of polarity in tip-growing cells is derived from studies of rhizoid emergence in zygotes of the brown alga *Fucus.*

Polarity of the *Fucus* Zygote is Initially Unstable but Later Fixed

Unlike higher plants, and even unlike most other algae, certain brown algae, particularly *Fucus* and *Pelvetia,* shed their eggs and sperm into the sea, where fertilization occurs. The zygote initially lacks a cell wall and is completely apolar.

However, a polar axis is established within a few hours of fertilization in response to unilateral light. The axis is not stable at first and can be changed many times simply by a change in the direction of the light. The polar axis becomes fixed within 10 to 14 hours after fertilization, after which it cannot be altered by a change in the direction of the light, and a structure that will become the rhizoid begins to protrude from the darkened side.

Even though the zygote is still a single cell, it is polarized, with two structurally different poles, the thallus and rhizoid poles, which have radically different fates. The first zygotic division is asymmetric and occurs at right angles to the polar axis. It establishes two cells, each having the identity of the pole from which it emerges. The smaller *rhizoid cell* produces a filament of cells and the holdfast that attaches the plant to rocks. The larger *thallus cell* divides to produce a globular mass of cells, from which the main body of the plant, the thallus, forms.

Calcium Channel Redistribution Leads to a Transcellular Current

A dye known as fluorescent dihydropyridine (FL-DHP), which binds to calcium channels in mammalian cells and can interfere with calcium uptake, has been used as a probe to examine the polarization of the *Fucus* zygote cell membrane.

During the first 6 to 8 hours after fertilization, and in response to unilateral light, FL-DHP receptors, presumably calcium channels, are redistributed and become localized on the dark side of the *Fucus* zygote at what will become the rhizoid pole. If the direction of the light is changed, the location of the FL-DHP–binding proteins also changes, again becoming concentrated on the dark side of the zygote.

One of the earliest events in the establishment of polarity after exposure of the zygotes to unilateral light is the induction of a positive current that flows into the cell on the shaded side and loops back through the external medium. This transcellular current, which is due largely to calcium ions, is on the order of 100 pA (1 picoampere = 10^{-12} ampere) and develops in advance of any morphological sign of polarity.

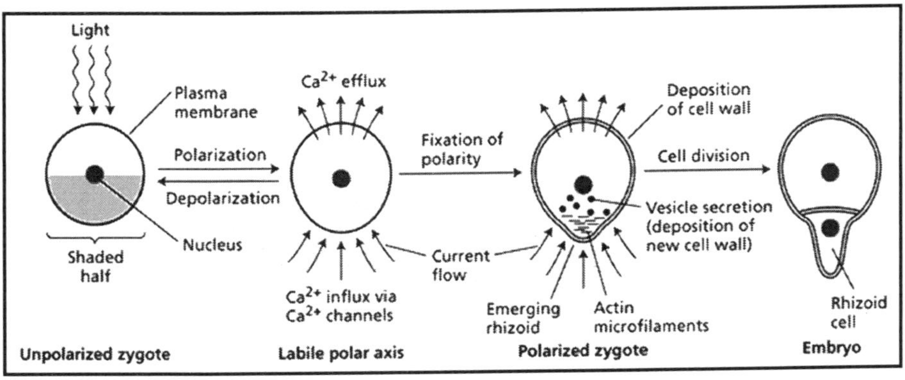

Fig. The Events Involved in the Establishment of Polarity in Zygotes of the brown alga *Fucus.*

(A) The zygote is polarized by an asymmetric stimulus from the environment, such as unilateral light. (B) A current flows (charged calcium ions move) through the polarized but still spherical zygote at the site at which the rhizoid will emerge, driven by Ca^{2+} uptake in the shaded half of the cell, from which the rhizoid will emerge. The current (Ca^{2+}) flows out on the opposite side.

(C) Cell polarity becomes fixed when actin microfilaments assemble at the site of rhizoid emergence and a cell wall is assembled around the zygote. The cell wall completely surrounds the zygote, but its composition differs in the rhizoid and thallus halves. Vesicles containing sulfated polysaccharides are transported by actin filaments to the plasma membrane and deposited in the cell wall only at the site of rhizoid emergence. (D) Finally, the zygote divides, and the rhizoid cell grows at its tip.

The relevance of this phenomenon to the establishment of cell polarity is indicated by the fact that the site at which positive charge flows into the cell is also the site of subsequent rhizoid outgrowth. Transcellular ionic currents also have been observed in some tip-growing cells of higher plants, including root hairs and germinating pollen grains, but they have not yet been detected in plant cells that expand by diffuse growth.

The ionic current in polarizing *Fucus* zygotes likely arises from the redistribution of ion channels in the plasma membrane. This current generates an electric-potential difference across the membrane that could alter other membrane properties, such as the distribution of membrane proteins.

The current also alters the ionic composition of the cytosol; the magnitude of the effect depends on the ions that form the current, which vary in different cell types. Because the Ca^{2+} concentration in the cytosol is usually very low (between 10^{-6} and 10^{-8} μM), even a small increase in intracellular Ca^{2+} concentration could have a marked effect.

Changes in Cytosolic Calcium Have Been Measured

The fluorescent calcium-indicator dye *fura*-2 has been used to detect intracellular cytosolic calcium gradients in both plant and animal cells. Since the excitation spectrum of fura-2 shifts when it binds calcium, we can determine the intracellular concentrations of both forms of the dye (with and without bound calcium) by exciting them with the appropriate two wavelengths.

The ratio of the two emissions provides a measure of the calcium concentration that is independent of the dye concentration. Using this technique to measure the cytosolic Ca^{2+} gradient in growing *Fucus* rhizoids, researchers observed a gradient from 450 nM at the tip to 100 nM at the base. This calcium ion gradient may be important for the polarized assembly of actin filaments and vectoral discharge of Golgi-derived vesicles at the rhizoid pole.

Calcium ion gradients also have been shown to be established in other tip-growing cells, such as germinating pollen grains, where the direction of growth and the region in which Golgi vesicles carrying cell wall components are deposited are predicted by the calcium gradient.

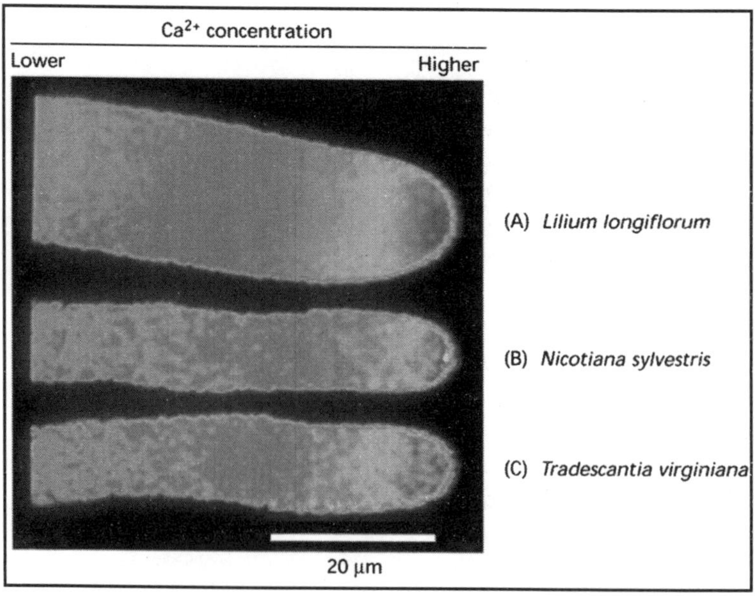

Fig. Tip-growing pollen tubes exhibit a steep intracellular gradient of calcium ion concentration, with the highest levels at the growing tip.

Pollen tubes from three species of plants were microinjected with a fluorescent calcium indicator dye to demonstrate this Ca^{2+} gradient, ranging from about 1 μM at the extreme tip to about 0.2 μM at the base.

The Actin Cytoskeleton Becomes Polarized

Inhibition of protein synthesis by treatment with the drug *cycloheximide* delays rhizoid formation but does not stop the establishment of polarity. This result suggests that although rhizoid formation requires the synthesis of new proteins, the establishment of polarity does not, so polarity development must involve a redistribution of existing components. Actin filaments become preferentially localized in the rhizoid half of the zygote and extend from the vicinity of the nucleus to the plasma membrane at the site marked by localized FL-DHP–binding proteins as polarity develops.

Actin filaments are dynamic, and their assembly from a pool of monomeric actin subunits is highly regulated. Actin filaments are localized in the rhizoid half of the cell presumably in response to a change in the assembly or disassembly dynamics, and this localization involves a redistribution of the cellular actin.

The fixation of polarity can be blocked by *cytochalasin B*, a drug that destroys actin filaments. Treatment with cytochalasin B also disrupts the ionic current that develops during polarization. Actin filaments participate in cell secretion by transporting secretory vesicles to the plasma membrane in many types of cells.

Secretion of Golgi-Derived Vesicles Is Localized

Cytoplasmic organelles become asymmetrically distributed before the polarity of the cell is fixed. Golgi-derived vesicles accumulate at the site of the future outgrowth in the rhizoid half of the *Fucus* zygote as polarization is induced by light. For several hours after the induction of polarity, the asymmetric distribution of vesicles can be changed by subsequent treatment with light from a different direction. During this phase when changes are possible, cytoplasmic vesicles accumulate at the prospective site of rhizoid formation on the shaded side of the cell. Subsequently, the vesicles fuse with the plasma membrane, secreting their contents into the cell wall, leading to rhizoid emergence. Evidence suggests that calcium gradients play a role in triggering vesicle discharge at the pole.

The fixation of light-induced polarity in *Fucus* zygotes also requires the presence of the cell wall. Removal of the wall with cell wall–digesting enzymes does not prevent the induction of polarity by unilateral light, but it does prevent fixation of the axis. The orientation of the embryonic axis remains unstable until a new cell wall is synthesized.

(A) A toluidine blue stained *Fucus* zygote before fixation of the polar axis. The toluidine blue dye stains a wall component in vesicles or in walls. (B) The embryo after establishment of rhizoid–thallus polarity, with toluidine blue staining confined to the cell wall in the rhizoid portion of the embryo.

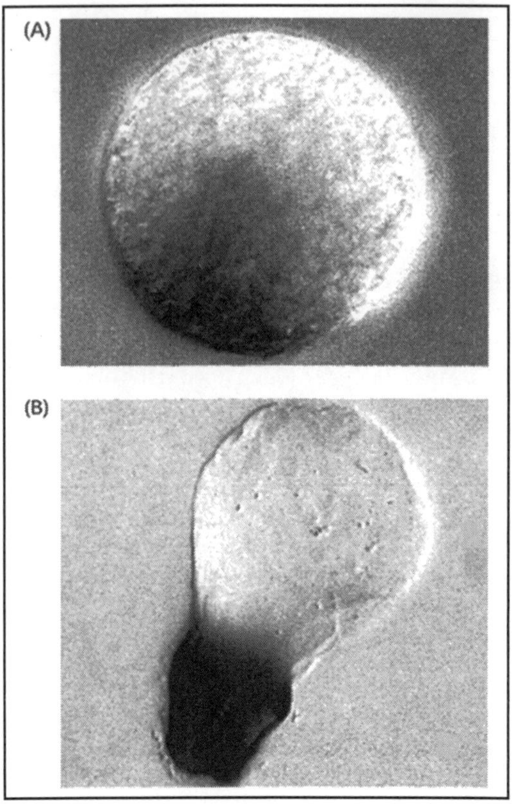

Fig. Cytoplasmic Vesicles that contain Specific cell wall Components become Asymmetrically Distributed early in the Development of *Fucus* zygote polarity.

From these observations it is possible to envisage a chain of events that might occur during the establishment of cell polarity in tip-growing cells.

- The ionic currents that arise early in the process may initiate the formation of an intracellular Ca^{2+} gradient.
- The increase in intracellular calcium at the tip may help organize and stabilize the microfilament components of the cytoskeleton.
- The cytoskeleton directs vesicle traffic to the growing tip.
- A calcium gradient, perhaps immediately adjacent to the plasma membrane, may promote vesicle fusion with the plasma membrane, although this remains to be demonstrated.
- Finally, cell wall deposition fixes the polarity of the cell, possibly by interacting with transmembrane proteins.

THE PREPROPHASE BAND OF MICROTUBULES

As we saw, microtubules in the cortical cytoplasm of growing plant cells are important in determining the orientation of cellulose microfibrils in the cell wall. When dividing cells enter the G_2 phase of the cell cycle, the interphase cortical microtubule array is replaced by a narrow band of interconnected microtubules and microfilaments encircling the nucleus. This circular array of microtubules and microfilaments is called the preprophase band.

The preprophase band was first observed in 1966 by Jeremy Pickett-Heaps and Donald Northcote at Cambridge University. It appears as a transient ring of cytoskeletal elements in the cortical cytoplasm, near the plasma membrane. Generally the preprophase band is 2 to 3 μm wide and contains from 10 to more than 100 microtubules, interspersed and interconnected with actin microfilaments.

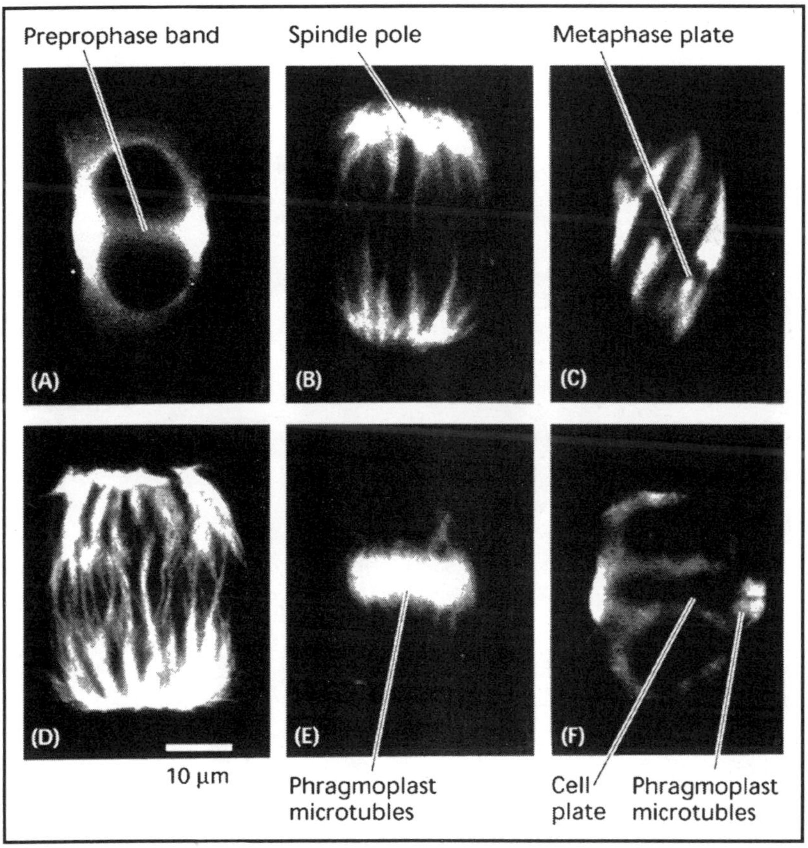

Fig. Immunofluorescent Staining of the Cytoskeleton in onion (*Allium cepa*) root tip cells with an Antibody to the Microtubule protein, tubulin.

As cells progress through their cell cycle, different cytoskeletal arrays that contain microtubules are formed. (A) A preprophase band appears after DNA replication is complete, but before the cell enters mitosis. It can be seen as two bright spots on either side of the unstained nucleus; the image of the band is fainter as it passes across the nucleus. (B) The mitotic spindle forms in late prophase, with spindle microtubules emanating from the polar regions to form the spindle fibers. (C) The spindle fibers hold the chromosomes (dark line) in the metaphase plate. In this cell the alignment of the plate is oblique, which is relatively unusual.

(D) The chromosomes move toward their poles within the array of spindle microtubules. (E) The phragmoplast forms in telophase and is seen here as a bright band in the centre of the cell that consists of microtubules, microfilaments, and vesicles in the early stage of cell plate formation. (F) Phragmoplast microtubules form an annular ring around the perimeter of the cell plate and appear only at the edges of the cell. Bar = 10 μm.

As the cell enters mitosis, microtubules begin to assemble at the surface of the nuclear envelope, and as the nuclear envelope breaks down, they invade the nuclear area to form the mitotic spindle. The preprophase band disappears as the spindle forms, leaving behind an actin-depleted zone. The plane in which cytokinesis occurs is determined before the cell initiates mitosis and is accurately predicted by the position of the preprophase band.

It has been suggested that the preprophase band determines the orientation of the mitotic spindle and that it marks the site at which the new cell wall will form during cytokinesis, but this has not been established. The position of the preprophase band and the plane of cytokinesis both may be determined by another mechanism, possibly sensitive positional information.

GENES REQUIRED FOR SHOOT APICAL MERISTEM FORMATION

Another class of transcription factors, encoded by CLASS III HD-ZIP class genes, has recently been shown to play essential roles in establishing the Shoot Apical Meristem (SAM). Mutants which lack activity of the previously discussed *REV* gene, as well as the related *PHABULOSA* and *PHAVOLUTA* genes, are unable to form an SAM.

As will be discussed later, these genes also promote establishment of adaxial tissues in the leaves. As meristems often form in close association with these tissues (*e.g.*, in the axils of leaves), models have been proposed in which the expression of CLASS III HD-ZIP genes promote what could be termed "central" cell identities.

In the stem, the "central" identity would include regions that contain apical initials, as opposed to the organogenic regions of the flanking peripheral zone. In the context of leaf development, this "central identity" would encompass the adaxial tissues, including the asymmetrically positioned xylem elements of the leaf as well as closely associated domains from which axillary meristems arise. This type of model receives further support from studies in which the overexpression of CLASS III HD-ZIP class genes causes SAMs to form in abnormal positions.

The regulation of CLASS III HD-ZIP genes is complex. Their polarized patterns of expression, and the mutant phenotypes that result from perturbing this expression, suggests that the normal patterning of the apical domain depends on radial patterning processes. As discussed earlier, the proteins encoded by these genes contain START domains, suggesting their activity is regulated by as yet undefined sterol type signaling molecules. The region that encodes the START domain is further distinguished by characteristics that suggest these genes are targets of microRNA (miRNA) mediated repression of transcript levels. An early clue to this mode of regulation was the highly conserved nature of the START domain region, which was apparent not just at the protein level, as might be expected of a ligand-binding region, but also at the nucleotide level, which suggested conserved interactions between nucleic acids.

This model has received additional support by the characterization of genes that encode miRNAs that are complementary to the conserved nucleotide sequence motif. The validity of the model is supported further by the effects of "silent" mutations, which alter the nucleic acid sequence of the motif without affecting its coding potential. Mutations of this type would prevent miRNA binding, and thus block the normal down-regulation transcripts in peripheral domains and lead to the observed dominant mutant phenotypes.

The regulation of transcript levels by miRNAs is not completely understood. In some cases, miRNAs mediate the formation of the so-called RNA Induced Silencing Complexes (RISC), which suppress the translation of target transcripts. An alternative type of posttranscriptional regulation has been documented in which the annealing of miRNAs to their mRNAs in the nucleus triggers transcript degradation. More recent data suggests that miRNAs may also mediate transcriptional silencing through chromatin-mediated reduction of gene transcription.

By any of these mechanisms (which can collectively be termed RNAi, for RNA inhibition), one would expect that genes involved in RNAi would play an important role in patterning processes. Indeed, this expectation is met by the mutant phenotypes of genes whose products mediate RNAi processes. These include the Arabidopsis *ARGONAUTE* and *ZWILLE* (*ZLL*) (also known as *PINHEAD*) genes, both of which affect tissue polarity and meristem formation.

Although it is unclear precisely how *ZLL* functions in the SAM, mutations that disrupt it function to block the accumulation of STM protein, but have no immediate effect on *STM* transcript levels. Further support for the idea that RNAi-dependent mechanisms are involved in normal SAM function is offered by the recent description of the genes required for normal shoot formation in rice, several of which have been shown to encode several distinct components of RNAi machinery.

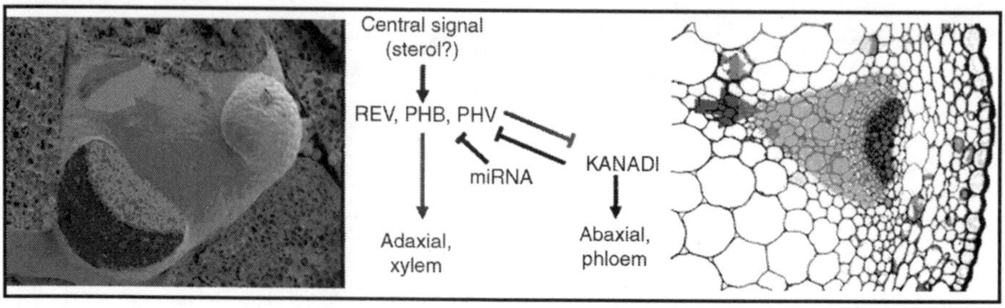

Fig. Model of how CLASS III HD-ZIP and KANADI activities pattern lateral organs and vasculature.

A centrally derived signal (red) activates CLASS III HD-ZIP genes, whose activity is antagonistic with that of KANADI activity. Both KANADI and MIR165/166 negatively regulate CLASS III HD-ZIP genes, although the relationship between the two is not presently known. In lateral organs, CLASS III HD-ZIP activity promotes adaxial fates and KANADI activity promotes abaxial fates, whereas in the vascular bundles, interactions between the two gene classes pattern the arrangement of xylem and phloem tissues. While the vascular bundle shown is already differentiated, the initial patterning events likely occur just below the apical meristem where provascular cells are being specified.

PLANT DEVELOPMENT DURING PHOTOSYNTHESIS

Photosynthesis is the process by which plants, some bacteria, and some protistans use the energy from sunlight to produce sugar, which cellular respiration converts into ATP, the "fuel" used by all living things. The conversion of unusable sunlight energy into usable chemical energy, is associated with the actions of the green pigment chlorophyll. Most of the time, the photosynthetic process uses water and releases the oxygen that we absolutely must have to stay alive. *The overall reaction of this process as*: Six molecules of water plus six molecules of carbon dioxide produce one molecule of sugar plus six molecules of oxygen.

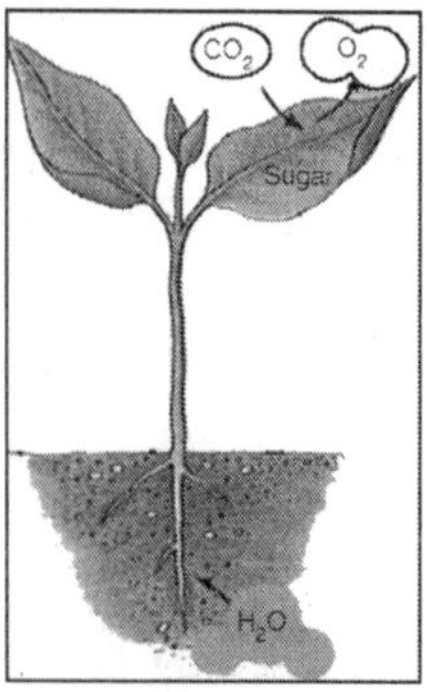

Fig. The Inputs and Outputs of the Photosynthetic Process.

Plants are the only photosynthetic organisms to have leaves (and not all plants have leaves). A leaf may be viewed as a solar collector crammed full of photosynthetic cells. The raw materials of photosynthesis, water and carbon dioxide, enter the cells of the leaf, and the products of photosynthesis, sugar and oxygen, leave the leaf.

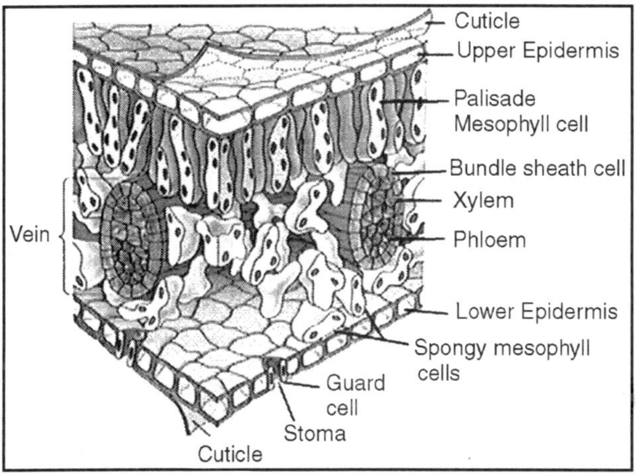

Fig. Cross Section of a leaf, Showing the Anatomical Features

Water enters the root and is transported up to the leaves through specialized plant cells known as xylem (pronounces zigh-lem). Land plants must guard against drying out (desiccation) and so have evolved specialized structures known as stomata to allow gas to enter and leave the leaf. Carbon dioxide cannot pass through the protective waxy layer covering the leaf (cuticle), but it can enter the leaf through an opening (the stoma; plural = stomata; Greek for hole) flanked by two guard cells.

Likewise, oxygen produced during photosynthesis can only pass out of the leaf through the opened stomata. Unfortunately for the plant, while these gases are moving between the inside and outside of the leaf, a great deal water is also lost. Cottonwood trees, will lose 100 gallons of water per hour during hot desert days. Carbon dioxide enters single-celled and aquatic autotrophs through no specialized structures.

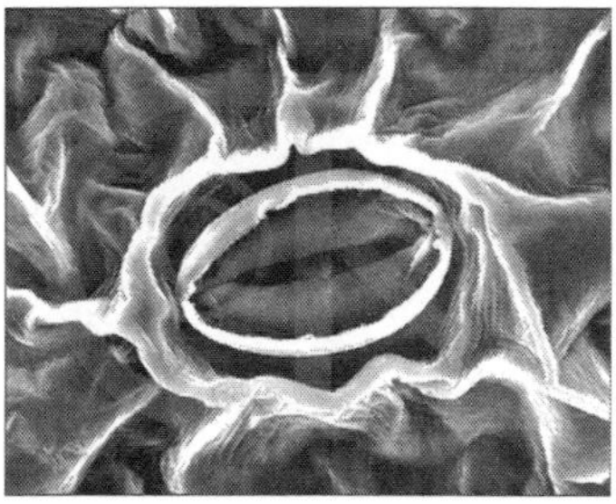

Fig. Pea Leaf Stoma, Vicea sp.

White light is separated into the different colours (=wavelengths) of light by passing it through a prism. Wavelength is defined as the distance from peak to peak (or trough to trough). The energy of is inversely porportional to the wavelength: longer wavelengths have less energy than do shorter ones.

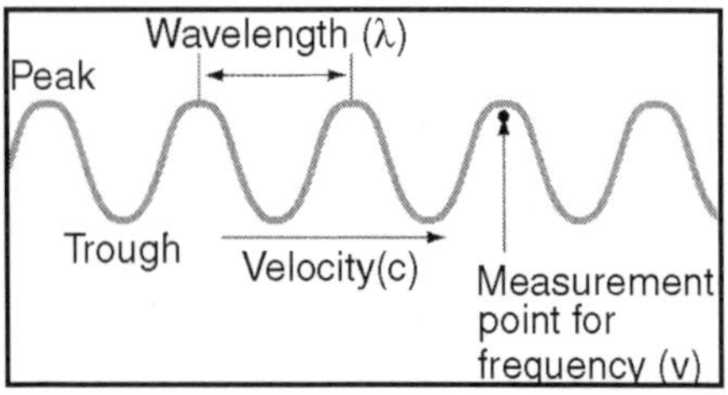

Fig. Wavelength and other Saspects of the Wave Nature of Light.

The order of colours is determined by the wavelength of light. Visible light is one small part of the electromagnetic spectrum. The longer the wavelength of visible light, the more red the colour. Likewise the shorter wavelengths are towards the violet side of the spectrum. Wavelengths longer than red are referred to as infrared, while those shorter than violet are ultraviolet.

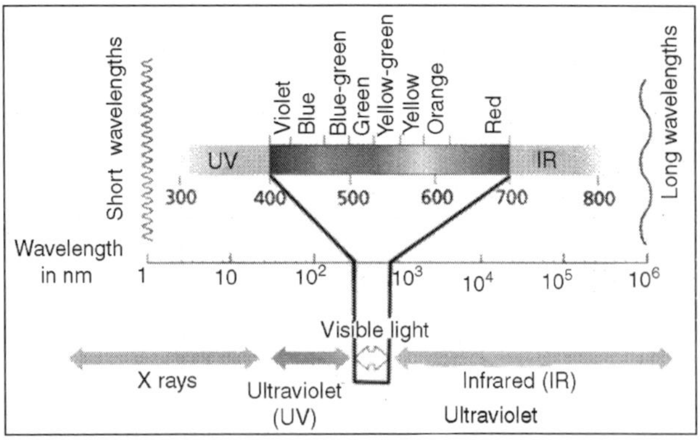

Fig. The Electromagnetic Spectrum.

Light behaves both as a wave and a particle. Wave properties of light include the bending of the wave path when passing from one material (medium) into another (*i.e.* the prism, rainbows, pencil in a glass-of-water, etc.). The particle properties are demonstrated by the photoelectric effect. Zinc exposed to ultraviolet light becomes positively charged because light energy forces electrons from the zinc.

These electrons can create an electrical current. Sodium, potassium and selenium have critical wavelengths in the visible light range. The critical wavelength is the maximum wavelength of light (visible or invisible) that creates a photoelectric effect.

ROLE OF CHLOROPHYLL AND ACCESSORY PIGMENTS

A pigment is any substance that absorbs light. The colour of the pigment comes from the wavelengths of light reflected (in other words, those not absorbed). Chlorophyll, the green pigment common to all photosynthetic cells, absorbs all wavelengths of visible light except green, which it reflects to be detected by our eyes.

Black pigments absorb all of the wavelengths that strike them. White pigments/lighter colours reflect all or almost all of the energy striking them. Pigments have their own characteristic absorption spectra, the absorption pattern of a given pigment.

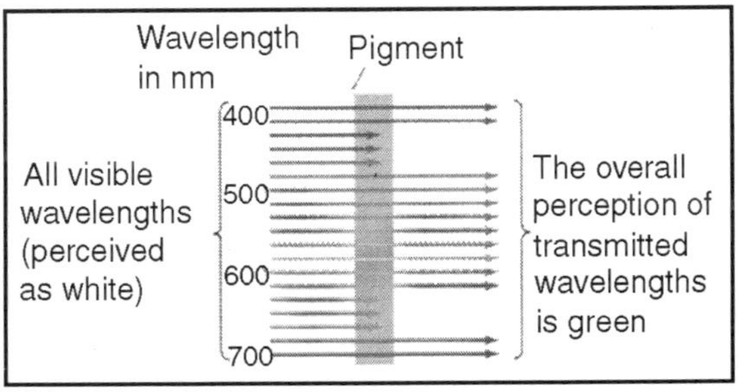

Fig. Absorption and Transmission of Different Wavelengths of light by a Hypothetical Pigment.

Chlorophyll is a complex molecule. Several modifications of chlorophyll occur among plants and other photosynthetic organisms. All photosynthetic organisms (plants, certain protistans, prochlorobacteria, and cyanobacteria) have chlorophyll a.

Accessory pigments absorb energy that chlorophyll a does not absorb. Accessory pigments include chlorophyll b (also c, d, and e in algae and protistans), xanthophylls, and carotenoids (such as beta-carotene). Chlorophyll a absorbs its energy from the Violet-Blue and Reddish orange-Red wavelengths, and little from the intermediate (Green-Yellow-Orange) wavelengths.

Carotenoids and chlorophyll b absorb some of the energy in the green wavelength. Why not so much in the orange and yellow wavelengths? Both chlorophylls also absorb in the orange-red end of the spectrum (with longer wavelengths and lower energy). The origins of photosynthetic organisms in the sea may account for this.

Shorter wavelengths (with more energy) do not penetrate much below 5 meters deep in sea water. The ability to absorb some energy from the longer (hence more penetrating) wavelengths might have been an advantage to early photosynthetic algae that were not able to be in the upper (photic) zone of the sea all the time.

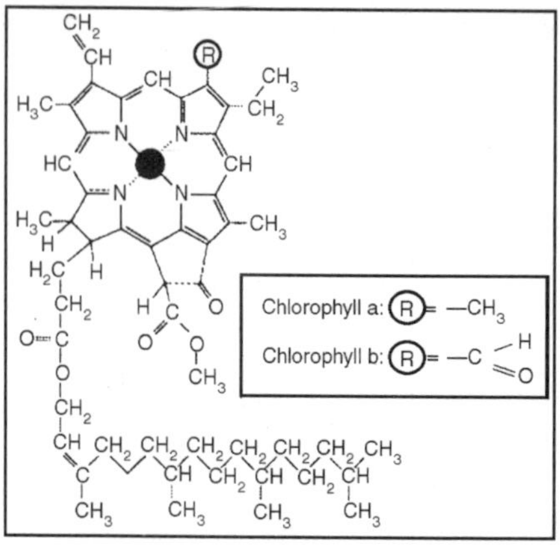

Fig. The Molecular Structure of Chlorophylls.

The action spectrum of photosynthesis is the relative effectiveness of different wavelengths of light at generating electrons. If a pigment absorbs light energy, one of three things will occur. Energy is dissipated as heat. The energy may be emitted immediately as a longer wavelength, a phenomenon known as fluorescence. Energy may trigger a chemical reaction, as in photosynthesis. Chlorophyll only triggers a chemical reaction when it is associated with proteins embedded in a membrane or the membrane infoldings found in photosynthetic prokaryotes such as cyanobacteria and prochlorobacteria.

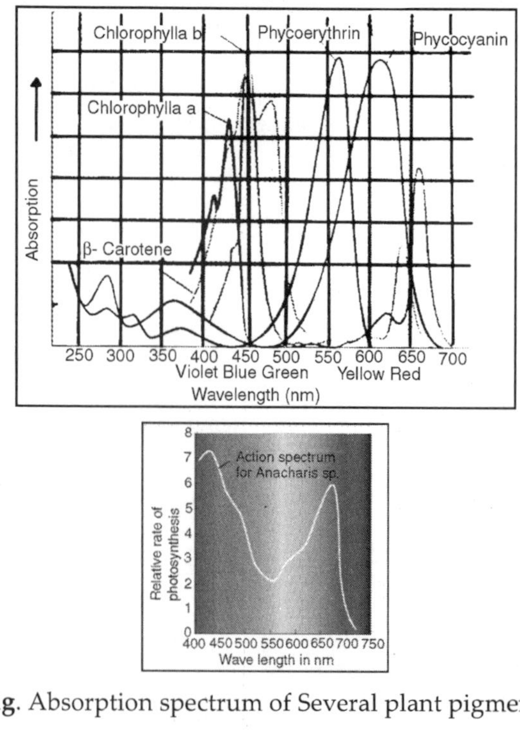

Fig. Absorption spectrum of Several plant pigments

ROLE OF THE CHLOROPLAST AND PHOTOSYNTHETIC MEMBRANES

The thylakoid is the structural unit of photosynthesis. Both photosynthetic prokaryotes and eukaryotes have these flattened sacs/vesicles containing photosynthetic chemicals. Only eukaryotes have chloroplasts with a surrounding membrane.

Thylakoids are stacked like pancakes in stacks known collectively as grana. The areas between grana are referred to as stroma. While the mitochondrion has two membrane systems, the chloroplast has three, forming three compartments.

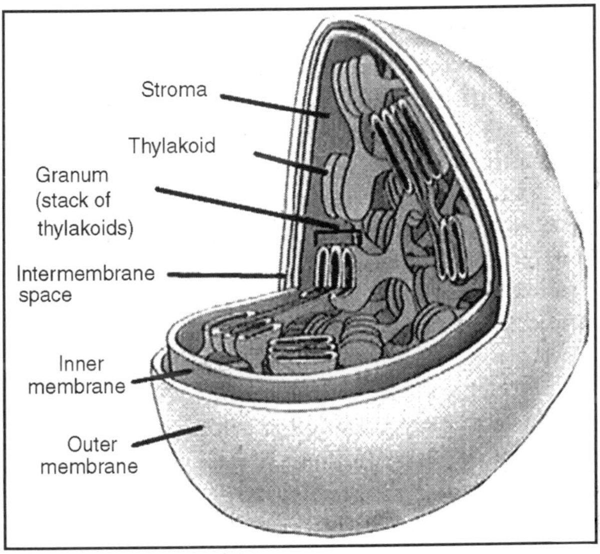

Fig. Structure of a Chloroplast.

STAGES OF PHOTOSYNTHESIS

Photosynthesis is a two stage process. The first process is the Light Dependent Process (Light Reactions), requires the direct energy of light to make energy carrier molecules that are used in the second process. The Light Independent Process (or Dark Reactions) occurs when the products of the Light Reaction are used to form C-C covalent bonds of carbohydrates. The Dark Reactions can usually occur in the dark, if the energy carriers from the light process are present.

Recent evidence suggests that a major enzyme of the Dark Reaction is indirectly stimulated by light, thus the term Dark Reaction is somewhat of a misnomer. The Light Reactions occur in the grana and the Dark Reactions take place in the stroma of the chloroplasts.

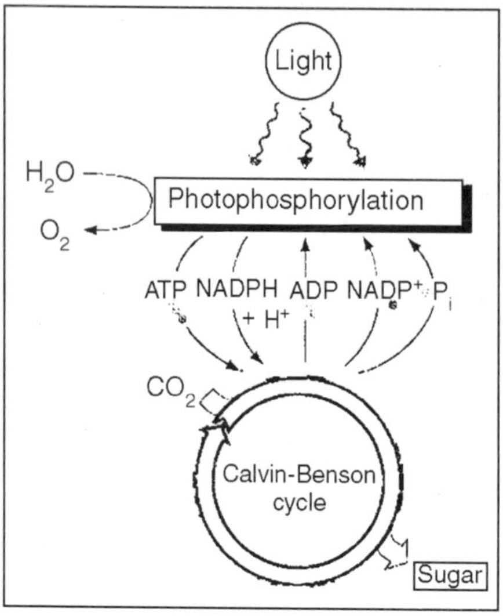

Fig. Overview of the two Steps in the Photosynthesis Process.

In the Light Dependent Processes (Light Reactions) light strikes chlorophyll a in such a way as to excite electrons to a higher energy state. In a series of reactions the energy is converted (along an electron transport process) into ATP and NADPH.

Water is split in the process, releasing oxygen as a by-product of the reaction. The ATP and NADPH are used to make C-C bonds in the Light Independent Process (Dark Reactions). In the Light Independent Process, carbon dioxide from the atmosphere (or water for aquatic/marine organisms) is captured and modified by the addition of Hydrogen to form carbohydrates (general formula of carbohydrates is $[CH_2O]_n$).

The incorporation of carbon dioxide into organic compounds is known as carbon fixation. The energy for this comes from the first phase of the photosynthetic process. Living systems cannot directly utilize light energy, but can, through a complicated series of reactions, convert it into C-C bond energy that can be released by glycolysis and other metabolic processes. Photosystems are arrangements of chlorophyll and other pigments packed into thylakoids.

Many Prokaryotes have only one photosystem, Photosystem II (so numbered because, while it was most likely the first to evolve, it was the second one discovered). Eukaryotes have Photosystem II plus Photosystem I. Photosystem I uses chlorophyll a, in the form referred to as P700. Photosystem II uses a form of chlorophyll a known as P680. Both "active" forms of chlorophyll a function in photosynthesis due to their association with proteins in the thylakoid membrane.

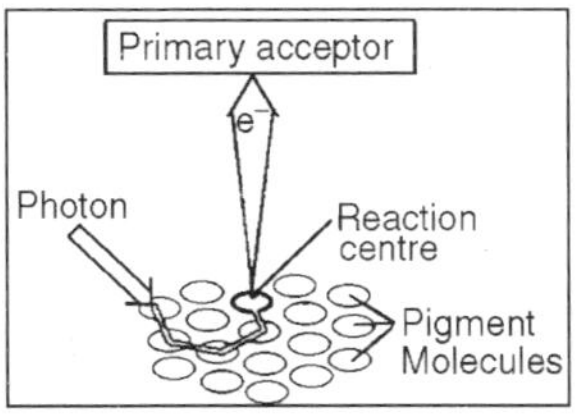

Fig. Action of a Photosystem

Photophosphorylation is the process of converting energy from a light-excited electron into the pyrophosphate bond of an ADP molecule. This occurs when the electrons from water are excited by the light in the presence of P680. The energy transfer is similar to the chemiosmotic electron transport occurring in the mitochondria. Light energy causes the removal of an electron from a molecule of P680 that is part of Photosystem II.

The P680 requires an electron, which is taken from a water molecule, breaking the water into H^+ ions and O^{-2} ions. These O^{-2} ions combine to form the diatomic O_2 that is released. The electron is "boosted" to a higher energy state and attached to a primary electron acceptor, which begins a series of redox reactions, passing the electron through a series of electron carriers, eventually attaching it to a molecule in Photosystem I. Light acts on a molecule of P700 in Photosystem I, causing an electron to be "boosted" to a still higher potential. The electron is attached to a different primary electron acceptor (that is a different molecule from the one associated with Photosystem II). The electron is passed again through a series of redox reactions, eventually being attached to $NADP^+$ and H^+ to form NADPH, an energy carrier needed in the Light Independent Reaction.

The electron from Photosystem II replaces the excited electron in the P700 molecule. There is thus a continuous flow of electrons from water to NADPH. This energy is used in Carbon Fixation. Cyclic Electron Flow occurs in some eukaryotes and primitive photosynthetic bacteria. No NADPH is produced, only ATP. This occurs when cells may require additional ATP, or when there is no $NADP^+$ to reduce to NADPH. In Photosystem II, the pumping to H ions into the thylakoid and the conversion of ADP^+ P into ATP is driven by electron gradients established in the thylakoid membrane.

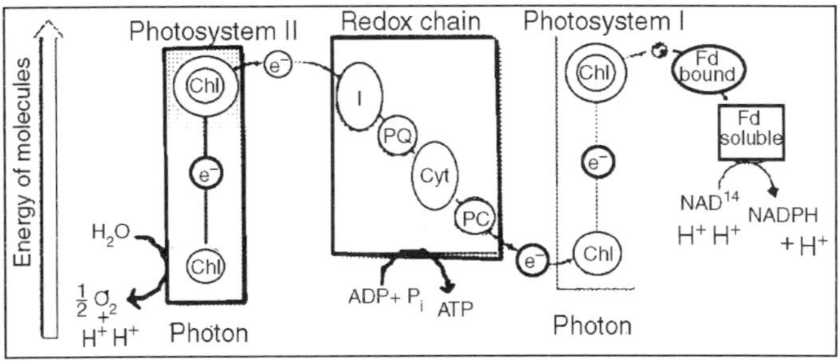

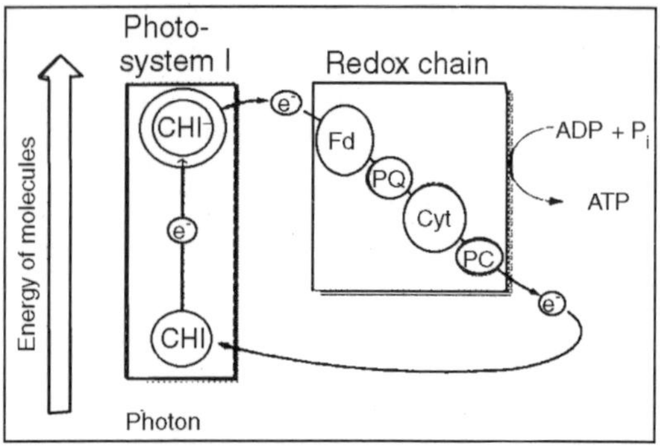

Fig. Noncyclic Photophosphorylation (top) and Cyclic Photophosphorylation (bottom). These processes are Better known as the light Reactions

Photophosphorylation: The process occurs in the chloroplast, and can link that to chemiosmotic synthesis of ATP.

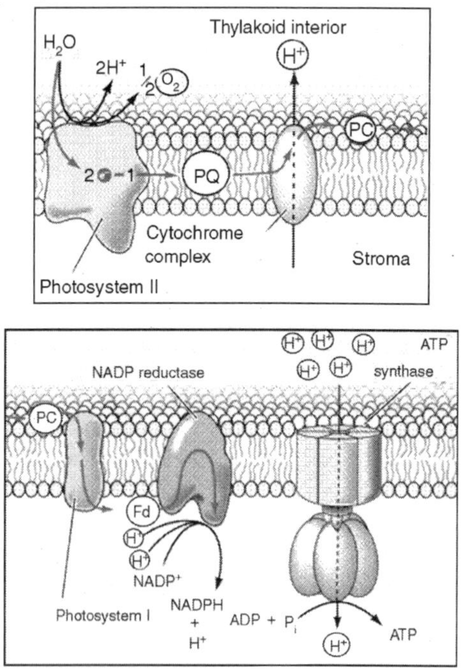

Fig. Chemiosmosis as it Operates in Photophosphorylation within a Chloroplast.

Halobacteria, which grow in extremely salty water, are facultative aerobes, they can grow when oxygen is absent. Purple pigments, known as retinal (a pigment also found in the human eye) act similar to chlorophyll. The complex of retinal and membrane proteins is known as bacteriorhodopsin, which generates electrons which establish a proton gradient that powers an

ADP-ATP pump, generating ATP from sunlight without chlorophyll. This supports the theory that chemiosmotic processes are universal in their ability to generate ATP. Some plants have developed a preliminary step to the Calvin Cycle (which is also referred to as a C-3 pathway), this preamble step is known as C-4. While most C-fixation begins with RuBP, C-4 begins with a new molecule, phosphoenolpyruvate (PEP), a 3-C chemical that is converted into oxaloacetic acid (OAA, a 4-C chemical) when carbon dioxide is combined with PEP.

The OAA is converted to Malic Acid and then transported from the mesophyll cell into the bundle-sheath cell, where OAA is broken down into PEP plus carbon dioxide. The carbon dioxide then enters the Calvin Cycle, with PEP returning to the mesophyll cell. The resulting sugars are now adjacent to the leaf veins and can readily be transported throughout the plant.

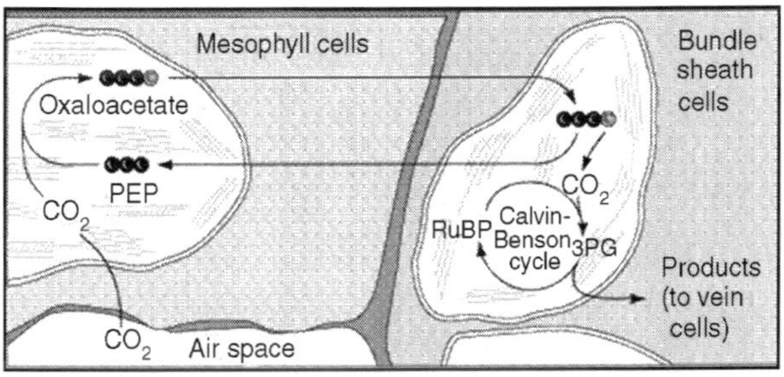

Fig. C-4 Photosynthsis Involves the Separation of Carbon Fixation and Carbohydrate Systhesis in space and time.

The capture of carbon dioxide by PEP is mediated by the enzyme PEP carboxylase, which has a stronger affinity for carbon dioxide than does RuBP carboxylase When carbon dioxide levels decline below the threshold for RuBP carboxylase, RuBP is catalyzed with oxygen instead of carbon dioxide.

The product of that reaction forms glycolic acid, a chemical that can be broken down by photorespiration, producing neither NADH nor ATP, in effect dismantling the Calvin Cycle. C-4 plants, which often grow close together, have had to adjust to decreased levels of carbon dioxide by artificially raising the carbon dioxide concentration in certain cells to prevent photorespiration.

C-4 plants evolved in the tropics and are adapted to higher temperatures than are the C-3 plants found at higher latitudes. Common C-4 plants include crabgrass, corn, and sugar cane. Note that OAA and Malic Acid also have functions in other processes, thus the chemicals would have been present in all plants, leading scientists to hypothesize that C-4 mechanisms evolved several times independently in response to a similar environmental condition, a type of evolution known as convergent evolution.

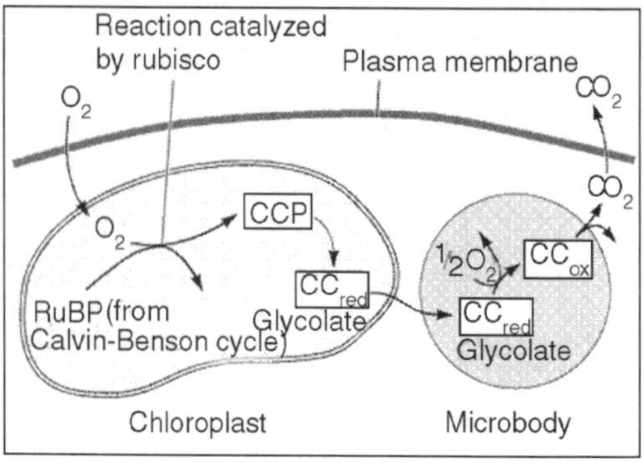

Fig. Photorespiration.

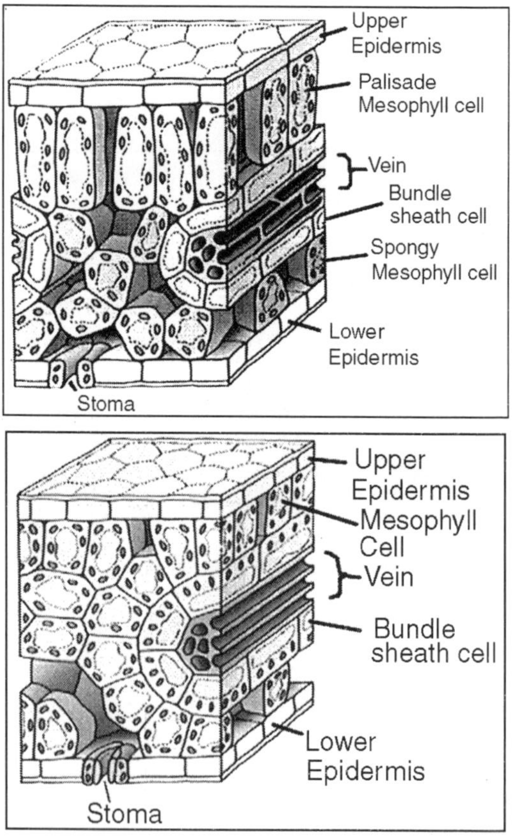

Fig. Leaf Anatomy of a C_3 (top) and C_4 (bottom) plant.

3

Plant Cells

Plant cells are eukaryotic cells, or cells with a membrane-bound nucleus. Unlike prokaryotic cells, the DNA in a plant cell is housed within the nucleus. In addition to having a nucleus, plant cells also contain other membrane-bound organelles, or tiny cellular structures, that carry out specific functions necessary for normal cellular operation.

Organelles have a wide range of responsibilities that include everything from producing hormones and enzymes to providing energy for a plant cell. Plant cells are similar to animal cells in that they are both eukaryotic cells and have similar organelles. Plant cells are generally larger than animal cells. While animal cells come in various sizes and tend to have irregular shapes, plant cells are more similar in size and are typically rectangular or cube shaped.

A plant cell also contains structures not found in an animal cell. Some of these include a cell wall, a large vacuole, and plastids. Plastids, such as chloroplasts, assist in storing and harvesting needed substances for the plant. Animal cells also contain structures such as centrioles, lysosomes, and cilia and flagella that are not typically found in plant cells.

CELL WALLS

- Surrounding the plasma membrane, the cell wall is a complex structure that provides protection and rigidity, regulates plant growth, and provides a porous medium for distribution of water and nutrients.

  The presence of a cell wall above all other characteristics distinguishes plant cells from animal cells. It presence is the basis of many of the characteristics of plants are organisms.

  Cell walls are not merely outer, inactive covering of the plant cell itself but are critical in function such as:

  - Prevents enlargement of the plant cell.
  - Also play important roles in the absorption, transport and secretion of substances in plants.

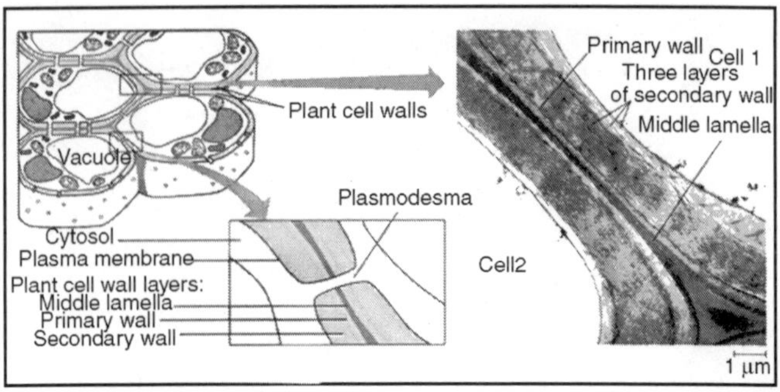

COMPONENTS OF THE CELL WALL

The most characteristic component of plant cell walls is cellulose, which largely determines their architecture. Cellulose is made of repeating molecules of glucose attached end to end in a b(1-4) linkage. These long thin cellulose molecules are united into a "Microfibril".

These microfibrils are arranged in a very regular, ordered arrangement and because of this the exhibit almost "crystalline" properties. These crystalline regions of the microfibrils are known as micelles. The microfibrils wind together to form fine threads which may coil around one another like a cable. Each "cable" is called a "Macrofibril".

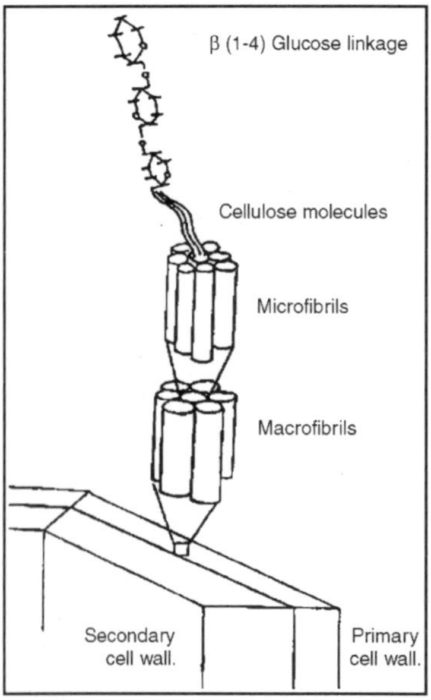

The cellulose framework of the wall is interpenetrated by a cross-linked "matrix" of non-cellulose molecules.

Some of these are polysaccharides called:

- Hemicellulloses

- Pectic substances *e.g.* "Pectin".

Other substances found in some plant cell walls are: Lignin - a very strong ridged molecule that is found in cells which have a strong supporting function.Suberin Cutin Waxes- Fatty substances which are commonly found in the cell walls of cells on the outside surface of the plant, and act to water proof the outer surface of the plant to prevent evaporation and dehydration.

THE EXTRACELLULAR MATRIX

Animal cells lack a rigid cell wall, but most secrete and are embedded in a sticky layer of glycoproteins called the "Extracellular Matrix". The Extracellular Matrix (ECM) is an elaborate covering outside animal cell membranes.

The major ingredients of the are glycoproteins secreted by the cells. The most abundant glycoprotein in the ECM is collagen which forms strong fibres outside the cell. This layer helps to hold cells together and can also have a protective and supportive function.

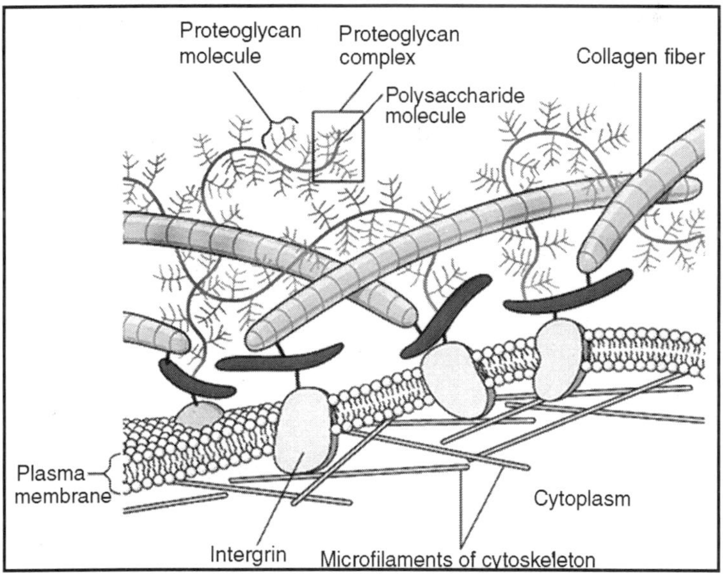

Fig. components of this Extracellular Matrix.

CELL JUNCTIONS

Plant cells are attached by their cell walls but are not isolated from one another. There are small pores called "Plasmodesmata" which provide a conection between the cytoplasm of one cell and the cell adjacent to it.

Tight junctions: these bind cells togetherforming a leakproof sheet. Such a sheet of tissue lines the digestive tract and prevents the contents from leaking into the surrounding tissues. Anchoring junctions: attaches adjacent cells to each other or to the extracellular matrix. Anchoring junctions rivet cells together with cytoskeletal fibres but still allow materials to pass along the spaces between cells.

Communicating junctions: are channels similar to the plasmodesmata in plant cells. They allow water and other small molecules to flow between neighbouring cells.

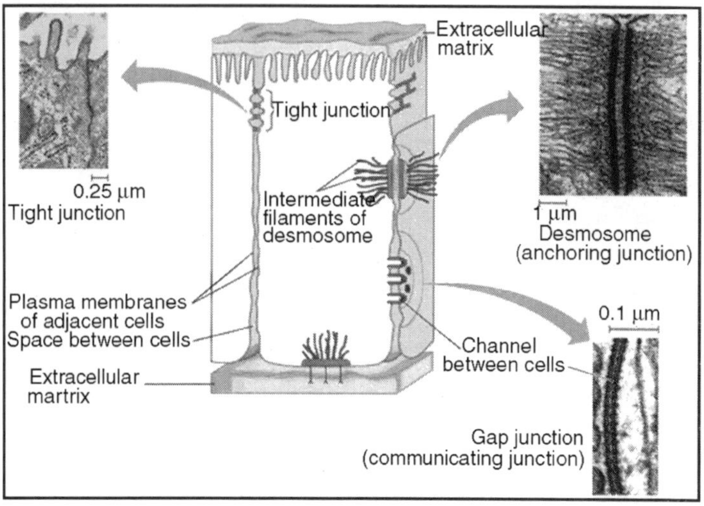

CELL ORGAN

CHLOROPLASTS

Millions of chloroplasts make leaves green. Plastids store and synthesize metabolic materials. Chloroplasts convert light into chemical energy, a process called photosynthesis—exclusive to plants.

VACUOLES

A large central sac, the vacuole, stores nutrients, waste products and pigments. It also provides structural support for the plant; filled with water, a vacuole may occupy 80 per cent or more of the cell and exert significant pressure against the cell wall.

PLASMODESMATA

Small tubes called plasmodesmata penetrate cell walls and connect neighbouring plant cells, to permit molecules to pass directly from one cell to another.

ENDOPLASMIC RETICULUM

The tubular network that connects the nucleus to everything else in the cell, the endoplasmic reticulum builds and moves compounds for use inside and outside the cell. It's common to both plants and animals; in plants, however, it also connects cell to cell through the plasmodesmata

CELL GROWTH AND VEGETATIVE DEVELOPMENT

ISOLATION OF MITOCHONDRIA

Although it is possible to measure respiration by intact tissues it is often useful to be able to isolate uncontaminated, intact and functional mitochondria. Here, we will consider some of the principles for isolating cell organelles. This discussion is not meant to be exhaustive nor will it be a cookbook with recipes to be followed step-by-step.

DISRUPTION OF THE TISSUE

The aim is to rupture as many of the cells in the tissue as possible in order to release the mitochondria, but without disrupting them. Whereas many mammalian tissues are easy to homogenize, plant tissues are much tougher mainly because of the cell walls. There are a number of standard homogenizers, including mixers with rotating knives or razor blades, juice extractors, ball mills, and mortar and pestle; the most suitable method for the specific plant material should be chosen. This may involve some initial optimization experiments. For special applications, cells may be disrupted by completely different methodology. Analyses of metabolites and cofactors in plant cell compartments require isolation of organelles (chloroplasts and mitochondria) on a subsecond timescale. For this purpose protoplasts devoid of cell wall are disrupted and organelles separated by rapid filtration.

HOMOGENIZATION MEDIUM

The medium for homogenizing rat liver only contains an osmoticum and a buffer. That will not work with plant tissues because of the disruption of the large central vacuole (absent in mammalian cells), which releases its fairly acidic content often containing phenols and other potentially harmful substances. Let us consider the components of the optimal homogenization medium for isolating intact and functional organelles from plant material, one by one.

- *Osmoticum*: This is required to prevent the mitochondria from losing their matrix content because of swelling and bursting. For the isolation of plant mitochondria, there is a tradition of using sucrose or mannitol at 0.2–0.5 M; whereas 0.15 M KCl is standard for the isolation of mammalian mitochondria. In neither case does this mimic precisely the conditions in the cytosol, but it works.
- *Buffer*: The pH in the cytosol of plant cells is around 7.5, and the homogenate should be buffered at pH 7–8. A buffer should therefore be included and a plethora of different buffers are available. Choose

an inexpensive one with a pKa of 7–8 (*e.g.*, 4-morpholinopropane sulfonic acid = MOPS) and be sure to use enough to counterbalance the acidic vacuolar content. The tissue type, the tissue/medium ratio, and the buffer concentration determine whether the acidic vacuolar content is properly neutralized. Obviously, special tissues like lemons or CAM plants harvested late at night or in the early morning, require special treatment!

- *Metal ion chelators*: EDTA is usually included to bind mainly Ca^{2+} and Mg^{2+}, which may otherwise activate proteolytic enzymes.
- *Antioxidants*: Antioxidants (reductants) such as ascorbic acid, dithiothreitol, or cysteine are used mainly to protect sulfhydryl groups on proteins against oxidation.
- *Bovine serum albumin (BSA)*: This protein from the blood of cows is usually included to protect the mitochondria: BSA binds fatty acids (its function in the blood) released by the action of phospholipases on phospholipids. BSA may also bind phenolics that are released from the vacuole. Both of these groups of compounds can otherwise be troublesome later. In addition, BSA may prevent proteases from degrading mitochondrial proteins by serving as an alternative substrate present at high concentrations.
- *Polyvinylpyrrolidone (PVP)*: A polymer that is often included, especially when using leaf material, to adsorb phenolic compounds.

FILTRATION AND PURIFICATION OF MITOCHONDRIA

To remove larger tissue fragments, cell walls, and often starch grains, the homogenate is always filtered, usually through a nylon net with a mesh size of 30-100 μm. The first step of the purification is usually differential centrifugation. This means that the filtered homogenate is centrifuged first at low speed for a short time (*e.g.*, 3000 g for 5–10 min.) and the pellet (containing cell wall fragments, starch grains, nuclei, intact plastids) is discarded. The supernatant is then centrifuged faster and for a longer time (*e.g.*, 10,000 g for 10–20 min.) and the pellet (termed *crude mitochondria*) is collected. It contains intact and damaged mitochondria as well as peroxisomes and plastid membranes. The supernatant is discarded.

As soon as the mitochondria have been pelleted and thus removed from potentially damaging soluble substances in the homogenate, the medium can be greatly simplified. Usually the so-called wash medium contains only osmoticum and weak buffer (no need for a strong buffer since there is no vacuole content to neutralize). However, it may be wise to include EDTA to bind divalent cations as well as BSA to adsorb any fatty acids released by the action of phospholipases. The resuspended crude mitochondria are often diluted to 50–100 ml with wash medium and subjected to another round of slow and fast centrifugation to remove more contaminants.

The aim of the next step is to separate the intact mitochondria from the contaminants. This usually involves a density gradient centrifugation with Percoll either alone or together with other dense materials. Percoll (colloidal silica coated with PVP) is inert and osmotically inactive. It has the further advantage that a Percoll gradient is self-generating because the Percoll microspheres pellet very slowly during the centrifugation. The crude mitochondria are layered on top of the wash medium containing the appropriate concentration of Percoll (usually 25–30%). If the Percoll concentration and the speed and duration of the centrifugation have been chosen correctly, the intact mitochondria will form a distinct band, well away from the contaminants. This band can be collected, washed free of Percoll by repeated dilution and centrifugation, and either used immediately, or frozen with 5% (v/v) dimethylsulfoxide (to preserve integrity) in liquid nitrogen and stored at–80°C until use. This preparation can be stable for months, retaining the ability to form a proton electrochemical gradient over the inner membrane.

The yield is 1–50 mg mitochondrial protein per 100 g fresh weight, depending on the tissue. This represents only a small fraction of the total amount of mitochondria in the tissue.

STRUCTURE OF PLANT CELL

To depict the structure of a plant cell, an epidermal cell of an onion will be used as an example. The epidermis is the final tissue that covers all organs above ground. The cells of the onion epidermis are common specimens on the first day of a German basic botanical course. Since they contain no chlorophyll, they are actually no "typical" plant cells. The picture above shows an onion's epidermal cells. They are elongated and the ratio of length to width can vary strongly. Each cell is enclosed by a wall. In the region of the cell poles and where three cells adjoin, large intercellular spaces can be observed.

Elsewhere a pectin-containing middle lamina cements neighbouring cells together like bricks. The cell wall is perforated at regular intervalls, so that adjoining cells are in contact. The holes of the perforation are called simple pits and the plasma cords that run through them plasmodesmata. The surface of the epidermal cells seems to be folded, an effect that is caused by the water-repellent, waxy cuticle.

PROTOPLASM, CYTOPLASM AND CYTOSOL: THE CELL'S CONTENT

The "living" content of a cell, the protoplasm, is surrounded by a membrane called plasma membrane or plasmalemma. The protoplasm is usually next to the cell wall, so that the plasmalemma can hardly be seen. To display it, the cells are transferred into a high salt or sugar solution. As a result the protoplasm shrinks and detaches itself from the wall. The process is reversible and is called plasmolysis.

This behaviour is due to the properties of the membrane and the plasma. It is reviewed in more detail elsewhere. A substance that causes plasmolysis is called plasmolyticum and - depending on its chemical composition (potassium ions orcalcium ions, for example)—the protoplasm takes on different shapes. The plasmolyticum has accordingly an influence on the properties of the membrane. The properties of the plasma membrane differ from that of the tonoplast. The tonoplast is the membrane that surrounds the vacuole. The difference is especially striking if cells with a coloured vacuole content are used. Often the vacuole is criss-crossed by numerous plasma cords. The plasma cannot therefore not simply be viewed as a solution that is influenced by the rules of hydrodynamics alone. Rather, it contains viscous, structure-determining components, whose chemical, physicochemical and structural properties have only been recognised recently and in fragments.

Cytoplasm and Caryoplasm

The nucleus is a rather conspicuous part of nearly every living plant cell. Its structure separated from the rest of the cell by the nuclear envelope, a membrane system that consists of two discrete membranes as can be seen on electromicroscopic images. The nuclear content is called the caryoplasm while that of the rest of the cell is called cytoplasm. But these terms are only valid at certain stages of a cell's life cycle. In the course of cell-division and mitosis, the nuclear envelope disintegrates and the nucleus is replaced by the chromosomes.

It makes consequently no sense to speak of caryo- and cytoplasm during these stages. The nucleus of plant cells is usually of a round or elliptic appearance, sometimes it is also shaped like a spindle. One nucleus per cell is the rule, but cells with two or more nuclei are no rare exception. The cells of certain algae of the genus *Chladophora* have many nuclei, they are polyenergid. The nucleoli that can often be perceived after staining are substructures of the nucleus. They, too, disintegrate during cell division and mitosis and do not reshape before a new nucleus has been formed.

Plastids

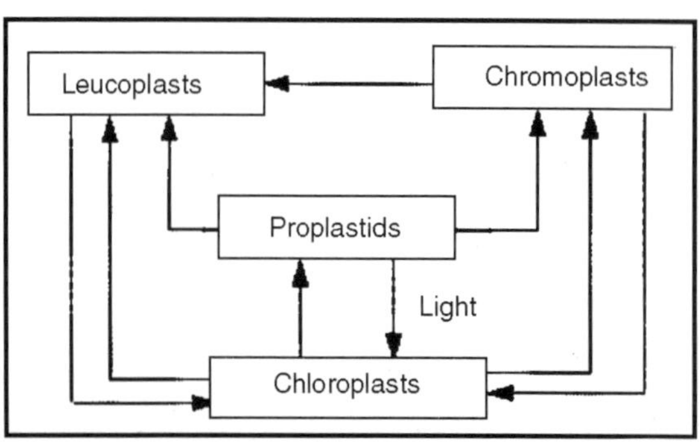

Plastids are organelles that occur only in plants. Their most prominent members are the chloroplasts. Others plastids are the coloured chromoplasts and the colourless leucoplasts as well as their proplastids. Proplastidsare vestigial bodies that are generated during germ cell development due to degeneration of plastids, for example. They may differentiate into complete plastids during the development of the plant embryo. Their ripening into chloroplasts occurs usually only after light exposure.

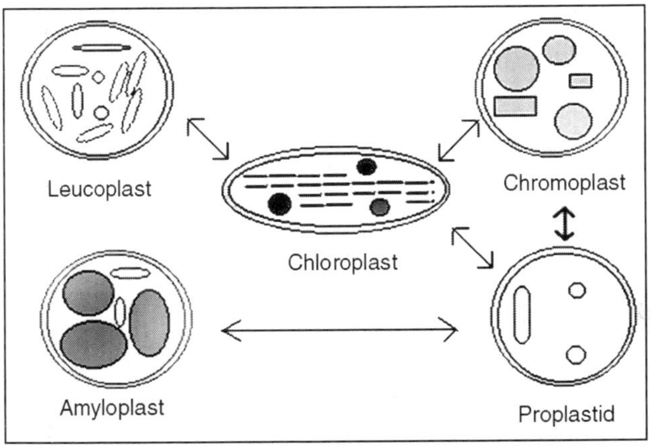

Chloroplasts contain the green plant colour chlorophyll. They are the places where photosynthesis takes place. Chloroplasts enable the plants to convert solar energy into chemical energy. Because of this process, plants are called primary producers. The existence of consumers, like most animals, depends on them. Chloroplasts occur in most cell types, but only in organs above ground.

They can be especially well observed in tissues consisting of a single layer as in the leaflike structures of some mosses (*Funaria hygrometrica* or *Mnium hornum*) or in the water plant *Vallisneria*). Here they are rather large and of a lens-shaped appearance. During daytime, when the light is diffuse, they occur mainly at the upper and lower surface of the chloroplast. They appear to be round under top view.

If exposed to strong light, they gather in parallel to the lateral sides of the cell which gives them an elleptic appearance upon top view. Chloroplasts are the site of starch production and -storage. Starch can easily be detected with the aid of potassium iodide (LUGOL's reagent). The starch-iodine complex is deeply blue-violet.

Starch production during photosynthesis can be made visible by placing a mask at a leaf that covers it partially while leaving some places exposed to sunlight. After one day of exposure, the leaf is first bleached to get rid of other pigments and afterwards treated with potassium iodide. An image is gained that is the exact replication of the mask and at the same time represents starch synthesis in the leaf. This experiment has first been done by J. v SACHS, probably the most outstanding plant physiologist of the 20th century. He thought that starch was the primary product of photosynthesis. This assumption proved wrong. It is well-known today that the first products of photosynthesis are monomeric sugars (glucose and others) and that only part of them is used for starch production.

The structures of the chloroplasts of higher cells resemble largely that of mosses. Their average diameter is 4 - 8 mm, an average cells contains 10 through 50. Their chlorophyll is unevenly distributed. At high resolutions chlorophyll-rich and chlorophyll-poor areas can be distinguished. This is due to the inner structure of the chloroplast: it is organised into grana (chlorophyll-rich) and stroma (chlorophyll-poor).

Upon stimulation with short-waved light (blue or violet) the chlorophyll emits an intensive red autofluorescence that looks especially impressive in a fluorescence microscope. The differences between grana and stroma become very obvious. The uniformity of the chloroplasts of all higher plants points out that the optimal form has been found rather early in evolution and has not been changed since.

This is different with algae. The chloroplasts of green algae (Chlorophyceens) are very varied in shape. Many species have just one chloroplast that covers nearly the whole space of the cell's interior. It is screw-like in *Spirogyra*-species, star-shaped in *Zygnema* and *Zygnemopsis* and netlike in *Oedogonium*. The disc-shaped chloroplast of *Mougeotia* can be viewed either from above or in profile depending on the amount of light used. Its rotation is a well-analysed example of an induced chloroplast movement. The chloroplasts of many species of algae contain often well-visible pyrenoids, structures, that produce and structure starch.

Chromoplasts are red, orange or yellow plastids. The colour is usually the result of yellow xantophyll and yellow to red carotinoids. Both compounds do also exist in chloroplasts, but are concealed by chlorophyll. Chlorophyll is broken down much faster than carotinoids as can be observed in the coloured leaves in autumn. Fluid transitions between chromo- and chloroplasts exist, just as between chromo- and leucoplasts. Typical chromoplasts cause the orange colour of the carrot, the red colour of the ripe pimento and tomato as well as the colour of numerous flowers.

Carotinoids are not very water-soluble and do therefore often crystallise within the chromoplasts. Their crystals can be disc-shaped, needle-like, jagged or sickle-like. In many cases, flower and leaf colours are caused by the coloured content of the vacuole. The colour of the vacuole and that of the plastids may lead to a mixed colour. The leaves of the copper beech, where the vacuole's content is red and that of the chloroplasts is green are a typical example. The plastids of the red and brown algae are traditionally counted among the chromoplasts although they contain chlorophyll. The green colour is concealed by the red phycoerythrin (Rhodophyceae) or the brown fucoxanthin (Phaeophyta). Leucoplasts are common, colourless plastids. They develop from proplastids, but form no homogeneous group of their own. A part of them can differentiate into chloroplasts or chromoplasts at light exposure, while this is not due for others. The guard cells, for example, contain leucoplasts, that are permanently exposed to light without developing into chloroplasts. Leucoplasts do also occur within colourless leaves (variegated leaves) or plant parts.

There exists a number of examples which show that they developed from chloroplasts that lost their ability to produce chlorophyll. There are even species, like *Neottia*, an orchid that cannot produce chlorophyll at all and are thus dependent on a parasitic or saprophytic lifestyle (saprophy is the feeding from dead organic material). A second class of leucoplasts occurs within the non-green tissues of otherwise green plants. It is especially common within roots. Though these leucoplasts are capable to become green, they do usually not since they are not exposed to light.

The leucoplasts of the calyptra (a calyptra is any hood or cap of cells protecting a plant part) contain starch and are therefore counted among the amyloplasts (starch-containing leucoplasts). They have, as is explained later, the function of statolithes, that have an important part in the perception of gravity (geotropism).

Starch

We got to know starch in the section above as a content of chloroplasts and leucoplasts (amyloplasts). It is produced by the polymerisation of glucose residues, which again are products of photosynthesis. Since the plant is able to transport sugars from leaf to root or from leaf to seed and fruit, starch production can also take place in these organs. Different species produce starch grains of different shape. Since the shape of starch grains informs about their origin, they are helpful in the identification of seeds and other starch-containing plant parts.

The following numbers show the variations in their diameters. Starch grains from potato tubers have a diameter of 70 - 100 μm, that of the endosperm of wheat 30 - 45 μm and that of corn endosperm 12 - 18 μm. Their shape reflects their development. The starch molecules are long-stretched and only sparsely branched. They are deposited within the plastids and their development begins at a so-called formation centre from where it proceeds radially. Layer follows layer and the thickness of one layer is dependent on the average molecular length.

A starch grain is therefore organised like a crystal (semi-crystalline). This can be shown very impressively with a polarisation microscope. Within the moistened specimen can a layering be viewed that is dependent on the water content of the single molecular parts.

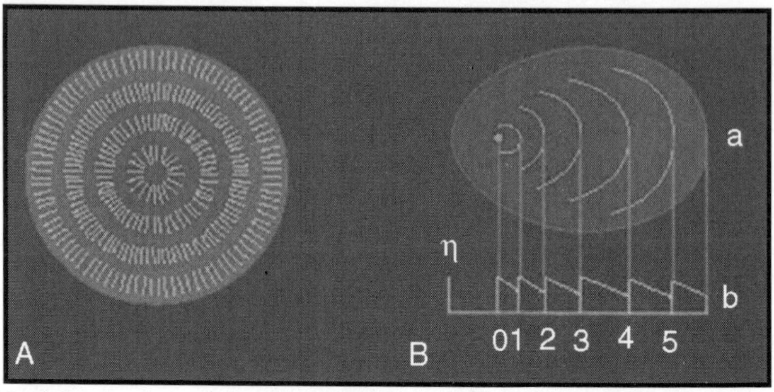

A. Model of a starch grain structure. The single lines symbolise starch molecules. They are arranged in a radial pattern. B. Layering of the starch grains, a. Formation centre and layer borders, b. Diagram of the refraction conditions. The ordinate shows the refraction index. (according to A. FREY-WYSSLING, 1938) The denser the molecular packaging, the less water is deposited. Layers with less water content refract light stronger than those with much water. After drying of the preparation no layering can be perceived any more. Depending the central or peripheral placing of the formation centre, starch grains with either concentric or eccentric layering develop. The starch grains of graminaceous plants (wheat, corn, etc.) are usually concentric, while those of the potato are always eccentric.

Sometimes plastids with two to three formation centres occur in graminaceous plants. This leads to the development of several starch grains. During growth common outer layers may be formed. Compound starch grains are typical for oat, they are built from a large number of smaller grains. Starch grains in bean seeds (*Phaseolus vulgaris*) are very big, their shape is round or oval, the spacing of the layers is very regular. Their centers can easily be hollowed out by addition of water displaying radial ruptures in microscopic images. In the sap of *Euphorbia splendens* dumb-bell shaped starch grains can be found.

CRYSTALLINE INCLUSIONS WITHIN CELLS

Many plant cells contain crystalline inclusions of different chemical composition and shape. Crystalline aggregations are called druses, bundles of needle-shaped crystals are termed raphids. Scanning electron microscopic images: Top picture: Calcium oxalate druse in the mesophyll cells of an oleander leaf *(Nerium oleander)*. Typical druse shape of dicots. Middle picture: Calcium oxalate needles (raphids) of a vanilla root (Orchidaceae). Typical raphid bundle of monocots. Lowest picture: Silicate bodies of silicate cells in the epidermis of *Schizachyrium sanguineum* (a gramineaen species of the old world tropics). Characteristic mineralisation of a gramineaen cell. [W. BARTHLOTT, MARTENS, 1979 (c), W. BARTHLOTT, unpublished (a, b)]

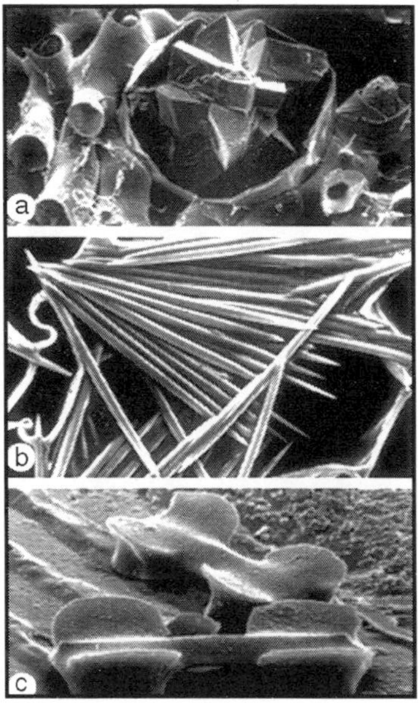

THE CELL WALL

Except for very few examples, plant cells are surrounded by a cellulose containing cell wall. It is flexible and distortable during growth, but loses its ability for distortion after growth has stopped, while a limited flexibility remains. Because of these changes, it is distinguished between primary and secondary cell walls.

As we will see when talking about electron microscopic pictures of the cell wall, both forms differ mainly in the arrangement of their cellulose microfibrils. While they are unorganised within the amorphous matrix of the primary cell wall, they are organised into several ordered layers that are arranged one on top of the other at right angles in the secondary cell wall. The secondary cell walls of many cells, especially those of vascular tissues, are incrusted with strengthening material.

Two important ones are:

1. Lignin, the ground substance of wood and
2. Suberin, the ground substance of cork

TYPES OF PLANT CELL

As a plant matures, its cells become specialised in order to perform certain functions necessary for survival. Some plant cells synthesize and store organic products, while others help to transport nutrients throughout the plant.

Some examples of specialised plant cell types include:

- Parenchyma Cells - although not highly specialised, these cells synthesize and store organic products in the plant.

- Collenchyma Cells - help to support plants while not restraining growth due to their lack of secondary walls and the absence of a hardening agent in their primary walls.
- Sclerenchyma Cells - provide a support function in plants, but unlike collenchyma cells, they have a hardening agent and are much more rigid.

Plant cells are grouped together into various tissues. These tissues can be simple, consisting of a single cell type, or complex, consisting of more than one cell type. Above and beyond tissues, plants also have a higher level of structure called plant tissue systems. There are three types of tissue systems: dermal tissue, vascular tissue, and ground tissue systems.

FUNCTION OF PLANT CELL WALLS

Higher plants resist the forces of gravity and powerful lateral forces through the cumulative strength of the walls that surround individual cells. These walls consist mainly of cellulose, noncellulosic polysaccharides and lignin, in proportions that depend upon the specific functions of the cell and its stage of development.

Spatially and temporally controlled heterogeneity in the physicochemical properties of wall polysaccharides is observed at the tissue and individual cell levels, and emerging *in situ* technologies are providing evidence that this heterogeneity also occurs across a single cell wall. We consider the origins of cell wall heterogeneity and identify contributing factors that are inherent in the molecular mechanisms of polysaccharide biosynthesis and are crucial for the changing biological functions of the wall during growth and development. We propose several key questions to be addressed in cell wall biology, together with an alternative two-phase model for the assembly of noncellulosic polysaccharides in plants.

Heterogeneity in the polysaccharide constituents of plant cell walls has been recognized since the earliest chemical analyses of isolated wall preparations. Clear and dramatic differences in average wall compositions are commonly observed, and actual differences are likely to be even greater than reported, because a good deal of heterogeneity can be disguised by the averaging effects of extracting constituent polysaccharides from plant or wall preparations that contain components from several different cell types. These observations are driving the development of a wide array of *in situ* methods for the definition of local wall composition, including methods based on enzymatic hydrolysis and mass spectroscopy, specific monoclonal antibodies, atomic force microscopy, Raman spectroscopy and real-time examination of cellular organelles.

In considering this heterogeneity, we shall see that certain aspects of the molecular mechanisms that mediate polysaccharide synthesis are inherently imprecise. In some cases the plant cell is able to control this imprecision and produces wall polysaccharides that are relatively homogeneous with respect to chemical structure and physicochemical properties. In other cases, the cell can generate polysaccharide products that are themselves chemically and physicochemically heterogeneous, a situation in which 'built-in' heterogeneity is a prerequisite for cell wall function.

In another example, a synthetic system with the same inherent limitations in precision builds polysaccharides that have both nonrandom and random structural features. The unusual and perhaps counterintuitive mechanism, through which an enzyme system can precisely generate random or other statistically defined structural features, is a characteristic of plant cell wall polysaccharide biosynthesis that appears to allow for both evolutionary and environmental diversity. And as if this were not complex enough already, there are postsynthetic modifications that can even further alter the chemical and physicochemical properties of wall polysaccharides.

DIVERSE FUNCTIONS OF CELL WALLS IN PLANTS

Plants do not have specialized skeletal systems of the type found in animals; instead, the considerable strength that enables large plants such as the giant *Sequoia* redwood tree to withstand the forces of gravity is provided by the cumulative strength of the walls that surround individual cells. But the load-bearing walls also have to be sufficiently flexible to withstand the large tensile and compressive forces that occur as, for example, the plant sways in the breeze.

The wall is subjected to further forces from within, where osmotic pressure represents another threat to cell and plant integrity. In young cells, the walls must be porous, to allow water, low–molecular weight nutrients and hormones to diffuse freely between cells and tissues, whereas in specialized vascular tissues required for long-distance movement of water and nutrients through the plant, the walls must be impermeable. In most tissues, cell walls are also important for intercellular adhesion. These various functional requirements are accommodated through the formation of reinforced gel structures consisting of cellulosic microfibrils, which have a high tensile strength, embedded in a gel-like matrix phase that consists predominantly of noncellulosic polysaccharides.

This matrix provides flexibility and some mechanical support for the cell, but its inherent porosity also allows water and other small molecules to diffuse through the wall. As cells of higher plants stop growing and increasingly assume a structural role, their walls thicken, cellulose microfibrils become laminated to increase wall strength and the walls often become encrusted with lignin, which is a large polyphenolic molecule that renders the wall resistant to compressive forces and restricts the passage of small molecules. At the same time, the wall will 'lock in' the final shape of the cell, and many different cell shapes can be detected in a single section of a plant organ. Smaller amounts of structural and enzymatic proteins and various proteoglycans and glycoproteins are also found in the wall, but these will not be discussed in detail here.

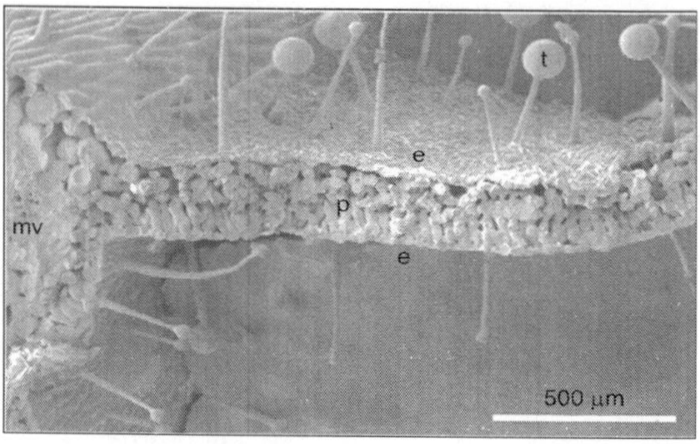

Fig. Heterogeneity in Cell Shape in a Leaf from *Nicotiana Benthamiana*.

In this mature leaf, which has stopped expanding, there are parenchyma cells (p) that have thin walls, fibre cells that have thickened walls, xylem cells that are no longer living and need a wall impervious to the water they transport, and trichome cells (t) that are greatly elongated and require a stiff wall to support them above the surface of the leaf. The cell walls surrounding these cell types play a major role in 'locking in' the final shape of the cell. The midvein (mv) and epidermal layers (e) are also indicated.

In addition to their crucial structural role, cell walls are likely to represent a significant source of metabolizable energy for some plants and may be subject to continual turnover in certain tissues as a result of this function. For example, the walls of the storage cells of seeds are not load bearing but in some cases are extraordinarily thick. Wall polysaccharides in these seeds act as an alternative to starch as a storage macromolecule. Examples include the galactomannans of seeds like fenugreek, carob, ivory nut and guar, as well as pectins in lupin seeds and xyloglucans in nasturtium and tamarind seeds. Grains of many grasses, particularly the economically important cereals such as wheat, maize, rice and barley, contain an endosperm that is rich in starch but also comprises a significant amount of cell wall material that can be used as a source of energy by the germinated grain. Indeed, it is estimated that up to 18% of metabolizable glucose in the germinated barley grain is released from the (1,3;1,4)-α-D-glucans of the starchy endosperm cell wall. A similar energy reserve function has been proposed for cell wall (1,3;1,4)-α-D-glucans in vegetative tissues of the grasses.

Finally, fragments of noncellulosic polysaccharides of cell walls have been implicated in growth regulation and in signaling pathways associated with a loss of cell wall integrity. Various oligogalacturonan fragments from pectic polysaccharides are bound by receptor-like kinases and can initiate a range of cellular changes, including defence responses that are executed following pathogen attack. Indeed, one might reasonably ask whether the evolution of the extreme complexity in pectic polysaccharides might have been driven by the potential of these polysaccharides to release an array of different signaling oligosaccharides. If this were true, an obvious next question would be whether other wall polysaccharides, such as the heteroxylans, undertake a similar function, especially in the *Poaceae,* where levels of pectic polysaccharides in the wall are very low.

Heterogeneity in cell Wall Polysaccharide Structures

Cellulose is the common component of most plant cell walls, and microfibrils of cellulose constitute the reinforcing rods of the wall. This macromolecule consists of unbranched, unsubstituted (1,4)-α-D-glucan chains that are able to form noncovalent microfibrillar complexes through extensive inter- and intramolecular hydrogen bonding and hydrophobic interactions. There is a good deal of heterogeneity in the chain length of cellulose, and its degree of crystallinity can also vary, but the fundamental chemical structure is essentially identical in the walls of different plants, tissues and organs. In contrast, the noncellulosic polysaccharides that compose the gel-like matrix of walls are structurally more complex and show considerable heterogeneity, from the tissue to the molecular level. Across the plant kingdom, a diverse array of polysaccharides is responsible for the formation of the wall matrix, but observed differences in the extant chemistries of the polysaccharides belie the diversity of the evolutionary histories that have converged to a common physicochemical outcome, namely the formation of a chemically heterogeneous but functionally competent gel-like component in the wall. In other words, the required physicochemical properties of wall polysaccharides can be met through a range of chemical strategies.

The backbone structures of these polysaccharides are based on (1,4)-β-linked monosaccharides in the case of cellulose, xyloglucan, (1,3;1,4)-β-glucan, heteroxylan and heteromannan, whereas the backbone of the galacturonans that are major constituents of pectic polysaccharides are based on (1,4)-α-linked chains of galacturonosyl residues. Extensive chain aggregation of the type found in cellulose is sterically hindered in the noncellulosic wall polysaccharides mainly through the addition of short oligosaccharides, monosaccharides or acetyl groups. However, in the (1,3;1,4)-β-glucans, main chain aggregation is prevented through the irregular conformation of the polysaccharide. Noncellulosic polysaccharides of walls from dicotyledonous plants are rich in pectic polysaccharides and xyloglucans, whereas walls in the *Poaceae* family of monocotyledons, which includes commercially important cereals and grasses, have relatively small amounts of pectic polysaccharides and xyloglucans and commensurately higher levels of heteroxylans and (1,3;1,4)-β-D-glucans. In the endosperm of legumes, walls often contain relatively high levels of mannans, galactomannans and galactoglucomannans.

Differences are also detected in the compositions of the 'primary' walls that are laid down during cell expansion and the 'secondary' walls that are deposited once cell expansion has ceased, although in practice it is often difficult to define when primary wall formation is replaced by secondary wall formation. Cellulose and lignin levels are typically higher in secondary walls than in primary walls. Compositional and physical differences have also been widely reported between primary and secondary walls; noncellulosic polysaccharides of secondary walls generally have lower degrees of branching or main chain substitution and are considered to be less soluble in aqueous media.

Thus, physical and chemical heterogeneity in plant cell walls at the species and developmental levels is extensive, but newer technologies have revealed additional levels of wall diversity. Monoclonal antibodies directed against cell wall polysaccharides specifically recognize epitopes that compose small regions of the polysaccharide and hence can distinguish different fine structural features in the polysaccharides. When sections of organs such as leaves and roots are probed with monoclonal antibodies directed against epitopes present on wall polysaccharides such as xyloglucans or pectins, binding varies by orders of magnitude between different cells in the section.

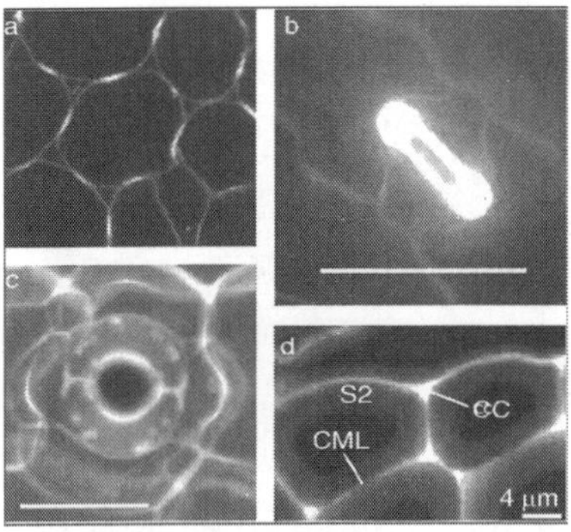

Fig. Heterogeneity in the Compositions of Cell Walls.

(a) Immunofluorescence microscopy and antibodies with different specificities for arabinan polysaccharides and their substructures were used to examine arabinan heterogeneity in stem tissues of *Arabidopsis*. In the pith parenchyma, the LM16 epitope was restricted to regions of adhered cell walls and was absent from cell wall regions surrounding intercellular air spaces.

(b) Distinctive UV autofluorescence associated with guard cells in maize. The data showed that guard cells had different amounts of phenolic esters of pectin than did epidermal cells.

(c) Heterogeneity in the patterns, colour and intensity of autofluorescence in the monocot *Commelina communis* has also been demonstrated.

(d) Raman images of a cross section of poplar wood prepared by integrating signals over defined wavenumber areas that showed constituents such as lignin and carbohydrate. The cell corner (CC) and the compound middle lamella (CML) adjacent to primary walls have a higher intensity of the lignin signal than in the S2 layer.

Molecular Mechanisms for Wall Polysaccharide Biosynthesis

The generally accepted mechanism of wall polysaccharide synthesis involves enzymes that add glycosyl residues from activated sugar nucleotide donors to the elongating polymeric chain. Two major classes of these glycosyltransferases have been recognized. One group, which is often referred to as the polysaccharide synthase group, is believed to catalyze the iterative addition of glycosyl residues to the end of the elongating polysaccharide chain, without releasing the chain between successive glycosylation events. It has been generally assumed that the incoming glycosyl residue is transferred from the sugar nucleotide donor substrate to the nonreducing end of the polysaccharide, although it has also been suggested that plant cell wall polysaccharide biosynthesis might involve addition of glycosyl residues to the reducing end of the elongating chain.

The iterative synthase enzymes are assumed to catalyze the polymerization not only of cellulose but also of the main chain of noncellulosic wall polysaccharides; these enzymes include the (1,4)-βmannan synthase that polymerizes the (1,4)-β-mannan backbone of mannans, galactomannans and galactoglucomannans and the (1,4)-β-glucan synthase that synthesizes the backbone of xyloglucans. These enzymes are encoded by members of the cellulose synthase (*CesA*) and cellulose synthase-like (*Csl*) gene superfamily and are members of the GT2 group of glycosyltransferases. The GAUT enzymes are also believed to be iterative glycosyltransferases that synthesize the (1,4)-β-galacturonan backbone of pectic polysaccharides, but the GAUT galacturonyl transferases belong to the GT8 group of enzymes, and their structures are likely to be quite distinct from the GT2 group involved in the synthesis of β-linked polysaccharide chains.

The second group of glycosyltransferases is believed to add single glycosyl substituents to the main chain residues or to add short oligosaccharide chains that are appended to the main chain through single catalytic encounters with the nascent polysaccharide chain. Examples of this group of glycosyltransferases include the arabinosyl transferases that add á-arabinofuranosyl or á-glucuronyl residues to the (1,4)-â-xylan backbone of arabinoxylans, the xylosyl transferases that add (1,6)-β-xylosyl residues to the (1,4)-β-glucan backbone of nascent xyloglucan molecules and the galactosyl transferases that add (1,6)-β-galactopyranosyl residues to the backbone of (1,4)-β-mannans or (1,4)-β-glucomannans. Despite the best efforts of biochemists over many decades, there are still a number of fundamental questions about the biosynthetic process that we have not answered. The mechanism of chain initiation is not known, and potential primer molecules for wall polysaccharide biosynthesis have not been convincingly identified. Similarly, the mechanisms controlling chain termination and thus the overall length of the polysaccharide chain have not been elucidated.

Do polysaccharide chains simply 'fall off' the synthase enzymes when they get to a certain length or when the enzyme itself begins to be turned over? Do the hydrolytic enzymes that have been increasingly recognized in tissues in which net synthesis of wall polysaccharides is occurring hydrolyze the elongating chain and release the nascent polysaccharide from the synthase enzyme? Might decreasing concentrations of glycosyl donor substrates lead to termination of the chain?

It was recently suggested that a terminal oligosaccharide detected at the reducing end of certain heteroxylans might be involved in chain termination. Whichever of these or other possibilities might apply, plant cell wall polysaccharide synthesis does not appear to be guided by a template, and it is difficult to envision how chain termination could be anything but an imprecise process. It follows that the resultant polysaccharide chain length will be heterogeneous, as is usually observed. This is not to say, however, that plant cells do not have some control over wall polysaccharide chain length, because clear differences are apparent, and in some cases chain lengths fall within reasonably narrow limits.

A second potential level of imprecision is afforded by the glycosyltransferases that add the various glycosyl units to the backbone of the wall polysaccharide. We do not know whether the addition of the substituent is a precise process or not, insofar as there are examples of relatively regular spacing of substituents along the chain and other examples in which substituents are, in statistical terms, asymmetrically distributed. In the former case, one might expect that the backbone synthase and the substituting glycosyltransferase would have to operate as a highly coordinated enzyme complex.

On the other hand, if the main chain synthase and glycosyltransferase were operating independently, it is again difficult to see how main chain substitution could be a precise process with respect to the extent and spacing of added substituents. Equally plausible is the possibility that an enzyme complex could mediate the generation of a polysaccharide in which every main chain residue was substituted or in which every second or third main chain residue was substituted.

This would result in a relatively homogeneous polysaccharide, but postsynthetic removal of substituents at regular intervals, as has been described for arabinoxylans, or asymmetrical postsynthetic modifications such as methyl esterification, as has been described for some pectic polysaccharides, could introduce heterogeneity into a polysaccharide that was initially homogeneous with respect to its fine structure. Against this background of heterogeneity that will arise generally as a result of biosynthetic mechanisms, the specific types of heterogeneity actually observed in the major groups of noncellulosic wall polysaccharides are summarized in the sections below.

Alternative Chemistries for Noncellulosic Polysaccharides

Two major chemical strategies have been used by plants to match noncellulosic polysaccharide structures to their functional requirements as components of the gel-like matrix phase of walls. Structurally regular polysaccharides with extended conformations of the type commonly found in cellulose and related wall polysaccharides will tend to align and aggregate in aqueous media; the aggregates will be stabilized through the formation of intermolecular hydrogen bonding and hydrophobic interactions between sugar rings.

This will reduce their solubility and enhance their ability to form gels. So, how do plants introduce limited and possibly controllable irregularity into the noncellulosic wall polysaccharides to limit this tendency to aggregate? In most instances the structurally regular and conformationally extended backbone or main chain that would normally align and aggregate into fibrillar or insoluble complexes is prevented from doing so through the addition of single monosaccharide substituents or short oligosaccharides along the otherwise regular backbone chain. During gel formation in the wall, the presence of some structurally regular 'junction zones' along the polysaccharide chain facilitates localized intermolecular interactions, which, in combination with 'solubilizing' noninteracting regions of the chain, allows the polysaccharide to form a gel.

An example of this chemical strategy is embodied in the xyloglucans, which consist of a (1,4)-β-glucan main chain. These 'cellulosic' backbone chains do not aggregate because they are substituted with short oligosaccharide chains at the C(O)6 of the main chain (1,4)-β-glucosyl residues. Similarly, the arabinoxylans of the *Poaceae* consist of a (1,4)-β-xylan main chain that would normally be insoluble and fibrillar in nature but is instead substituted with α-arabinosyl residues. There are also regions of the (1,4)-β-xylan backbone that carry no substituents; these will therefore be capable of intermolecular interactions that will result in limited junction zone formation and will allow gel formation in the wall.

The chemical structures of galactomannans and pectic polysaccharides reflect the same strategy. The second, quite different strategy to achieve wall polysaccharides with extended conformations but with limited tendencies to align and aggregate is exemplified by the relatively recently evolved (1,3;1,4)-β-glucans of the *Poaceae*. Like the xyloglucans, their structure is based upon a conformationally regular 'cellulosic' (1,4)-β-glucan backbone. However, in this alternative strategy, intermolecular aggregation is limited through the introduction of irregularly spaced (1,3)-β-linkages, which result in molecular 'kinks' in the otherwise 'cellulosic' backbone. Provided the (1,3)-β-linkages are spaced at irregular intervals, the asymmetric polysaccharide chains will not aggregate over extended regions.

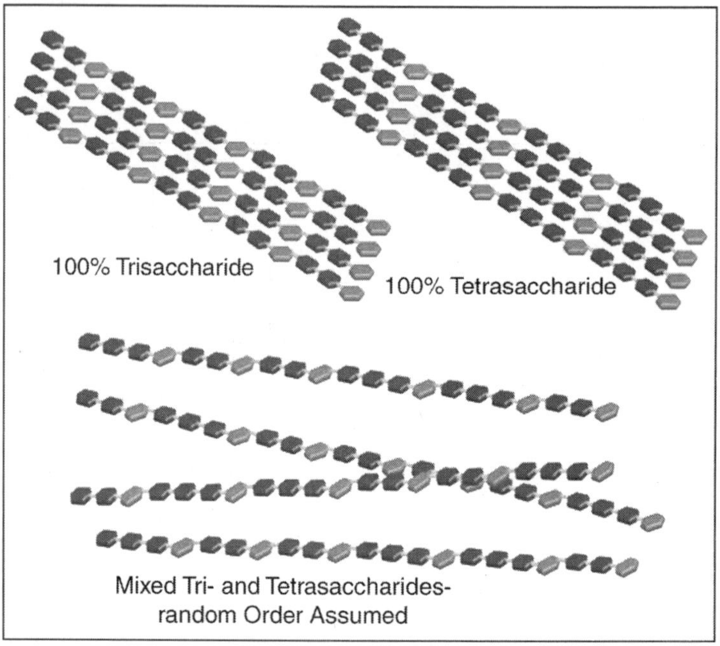

Fig. The Effects of (1,3;1,4)-β-glucan fine structure on aggregation and solubility.

The (1,3;1,4)-β-glucans of the *Poaceae* have structures based upon a conformationally regular 'cellulosic' (1,4)-β-glucan backbone. Intermolecular aggregation is limited through the introduction of irregularly spaced (1,3)-β-linkages, which result in molecular 'kinks' in the (1,4)-β-glucan backbone. The (1,3)-linked â-glucosyl residues are shown in red, whereas the (1,4)-â-glucosyl residues are shown in blue. Relatively high or relatively low ratios of cellotriosyl/cellotetraosyl units indicate a predominance of cellotriosyl or cellotetraosyl residues, respectively. If one or another of the oligosaccharide units predominates, the overall polysaccharide chain becomes more regular in shape and hence less soluble because of chain alignment. Random arrangement of a mixture of the trisaccharide and the tetrasaccharide units means that the chains will not aggregate over extended regions and can remain in aqueous solution, as depicted in the lower diagram. Such differences in physicochemical properties could be used to the plant's advantage in meeting different functional requirements in different cells or at different stages of development.

Heterogeneity in Pectic Polysaccharides

Pectins are a major component of the primary cell wall, particularly of dicotyledonous plants. Pectins contribute to the mechanical strength, porosity, adhesion and stiffness of the cell wall and have been implicated in many processes, including wall slippage, extension and intercellular signaling. In addition, pectic polysaccharides are a source of signaling molecules. Pectins are extremely complex molecules that include a number of structural classes, such as homogalacturonan, xylogalacturonan, apiogalacturonan and rhamnogalacturonan.

The galacturonan chains can chelate calcium ions, apiosyl residues can coordinate borate ions, and there is good evidence that covalent linkages form between different structural classes of the pectic polysaccharide group. Methylated and acetylated residues are often found on pectin and indeed on several other polysaccharides and can profoundly influence their structures and physicochemical properties. Considering this complexity, it is not unexpected that the level of heterogeneity in the pectic polysaccharides is high. Here we will explain just one example of how structural heterogeneity is generated and how this alters the physicochemical and functional properties of the cell wall matrix.

A common mechanism facilitating local changes in the properties of pectic polysaccharides is the degree to which their polygalacturonate residues are methyl esterified. This involves heterogeneity both within a chain and between chains and can now be characterized analytically. Methyl esterification will neutralize the charge on galacturonyl residues and thereby abolish the ability of the negatively charged galacturonyl residues to interact with calcium ions. Specific monoclonal antibodies, which can differentially detect pectins with particular methylation profiles, show spatial heterogeneity around a single higher-plant cell with respect to the pattern of pectin esterification, which may be linked to cell adhesion.

There are a number of pectin methyl esterases that selectively remove methyl groups from homogalacturonan in either a blockwise or a nonblockwise fashion. Where methyl groups are removed in a blockwise manner, an exposed region of the polysaccharide that is able to form calcium-mediated cross-links is created; these cross-links contribute to the 'stickiness' required for cell adhesion. Where the removal of ester groups is nonblockwise, the homogalacturonan appears to be less 'sticky', and such junction zones cannot form so readily. As one might expect, this particular form of homogalacturonan is found in regions where cells separate.

Heterogeneity in Heteroxylans

Heteroxylans have a (1,4)-β-xylan backbone that can be substituted with single á-arabinofuranosyl, á-glucuronosyl or 4-methyl-á-glucuronosyl residues. These residues can be linked to the $C(O)_3$ or $C(O)_2$ atoms of the backbone xylopyranosyl residues, or in some cases the xylopyranosyl residues can be substituted at both $C(O)_2$ and $C(O)_3$. Short oligosaccharide substituents are also found on some heteroxylans and, in the grasses in particular, á-arabinofuranosyl substituents can be esterified with hydroxycinnamic acids such as ferulic acid and *p*-coumaric acid.

Two carbohydrate-binding modules and two monoclonal antibodies have been used to locate specific types of xylans both within intact cell walls and in solubilized wall material. Both classes of probe revealed distinct types of heteroxylans, which differ in both structure and charge and are present in both primary and secondary walls. Similarly, in the endosperm of the grasses, wall arabinoxylans digested with (1,4)-β-xylan endohydrolase release a population of oligomeric arabinoxylosides that provide an enzymatic fingerprint of the polysaccharide's chemical structure. Examination of the arabinoxylan fingerprint from wheat endosperm walls showed variation in the structure and amount of this polysaccharide both spatially and temporally through grain development.

The results were confirmed through Fourier transform infrared microspectroscopy (FT-IR), which showed that arabinoxylans in developing grain contained more backbone xylosyl residues disubstituted with arabinosyl residues than did arabinoxylans in mature grain and that the structure of the arabinoxylan was dependent upon its position in either prismatic or central cells.

Heterogeneity in Xyloglucans

Xyloglucans are thought to be particularly important in the primary walls of many plants because of their ability to noncovalently cross-link cellulose microfibrils. Their structure is based on a cellulosic (1,4)-β-glucan backbone substituted at $C(O)_6$ with á-linked xylosyl residues, typically at about 70% of the backbone glucosyl residues. However, further substitution of xylosyl residues is required to prevent xyloglucan self-association, which interferes with binding to cellulose. A typical basic substructure involves xylosyl substituents on three consecutive glucosyl backbone residues, followed by an unsubstituted glucosyl residue.

The availability of endoglucanase enzymes that depolymerize the xyloglucan chain at unsubstituted glucosyl residues allows the isolation of oligosaccharides for detailed analysis. A wide range of carbohydrate and selective acetyl substituents on the basic xyloglucan structure have been characterized in this way. Most studies of partial glucanase digests are consistent with an essentially random arrangement of oligosaccharide substructures, but this may not always be the case. One promising new approach to characterizing the heterogeneity of xyloglucans is the use of mass spectrometry to define glucanase-derived oligosaccharides on very small sample sizes. The selectivity of specific endoglucanase digestion, coupled with the sensitivity of mass spectrometric detection, allows not only the profiling of small amounts of isolated xyloglucan polymers but also the profiling of xyloglucans present within specific cell types or in isolated organelle fractions. Furthermore, *in situ* analysis of xyloglucan oligosaccharide substructures in walls can be carried out at high resolution by aiming the ionization laser of a MALDI-TOF mass spectrometer at desired locations within enzyme-treated tissues.

Although mass spectral methods depend to some extent on prior knowledge of xyloglucan structure and cannot be easily applied to polysaccharides such as arabinoxylans, which consist only of pentose sugars, the method has potential to open up the dissection of location-specific heterogeneity in xyloglucan structures, especially if the method can be complemented with antibodies directed against an array of xyloglucan substructures.

Heterogeneity in (1,3;1,4)-β-glucans

The (1,3;1,4)-β-glucans, which are abundant in walls of the *Poaceae,* represent one example in which heterogeneity in fine structure appears to be essential for function in the wall and is therefore built into the (1,3;1,4)-β-glucan during biosynthesis. Detailed structural analyses of the (1,3;1,4)-β-glucans demonstrate that randomness can be recognized at one level, whereas other structural features of the molecule are nonrandom. How can this be?

The water-soluble (1,3;1,4)-β-glucan from barley endosperm consists of about 72% (1,4)-β-glucosyl residues and 28% (1,3)-β-glucosyl residues. The (1,3)-β-glucosyl residues are always flanked on both their reducing end and nonreducing end sides by (1,4)-β-glucosyl residues; two or more adjacent (1,3)-β-glucosyl residues are not found. In this respect the fine structure is nonrandom, because if the (1,3)-β-glucosyl residues and (1,4)-β-glucosyl residues were randomly distributed, on statistical grounds one would expect to find significant numbers of two or more adjacent (1,3)-β-glucosyl residues.

The (1,4)-β-linkages of most (1,3;1,4)-β-glucans are arranged predominantly in groups of two or three adjacent linkages. In the water-soluble (1,3;1,4)-β-glucan from barley grain, this arrangement accounts for about 90% of the polysaccharide. The remaining 10% consists of longer blocks of from 4 up to 12–15 adjacent (1,4)-β-glucosyl residues. If we focus on the groups of two and three adjacent (1,4)-β-glucosyl linkages, (1,3;1,4)-β-glucans can be viewed as (1,3)-β-linked copolymers of cellotriosyl and cellotetraosyl residues in which the ratio of cellotriosyl (DP3) to cellotetraosyl (DP4) units varies from 1.5 to 4.5, depending on the source of the (1,3;1,4)-β-glucan. Markov chain analyses were used to investigate the arrangement of the cellotriosyl (DP3) to cellotetraosyl (DP4) units and showed that the cellotriosyl

(DP3) to cellotetraosyl (DP4) units were arranged at random along the (1,3;1,4)-β-glucan chain.

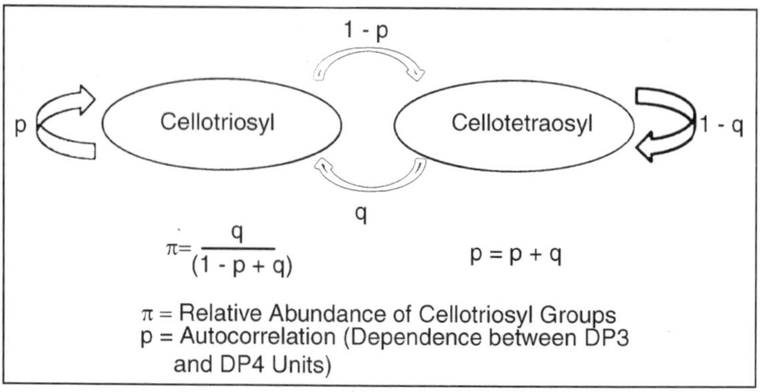

Fig. Fine Structural Analysis of the (1,3;1,4)-β-glucan from Barley.

The structure of a water-soluble barley (1,3;1,4)-β-glucan was investigated by treating it as a Markov chain. The polysaccharide consists mostly of cellotriosyl and cellotetraosyl units linked by (1,3)-β-linkages. This analysis was used to explore whether the addition of a particular unit oligosaccharide structure to the elongating chain depended on the adjacent unit oligosaccharide in the chain.

Through controlled enzymatic hydrolysis, such that most of the products represented two units of the polysaccharide chain, it was possible to quantitate the amount of hexasaccharide (which consisted of two cellotriosyl units), heptasaccharide (which consisted of one cellotriosyl unit and one cellotetraosyl unit) and octasaccharide (which consisted of two cellotetraosyl units). Applying this quantitative measure of near-neighbour analysis in the Markov chain would give a value for ρ of about 1.0 for blocks of adjacent cellotriosyl units and adjacent cellotetraosyl units, with values near –1.0 if the polysaccharide consisted of alternating cellotriosyl and cellotetraosyl units and values close to zero if the cellotriosyl and cellotetraosyl units were randomly arranged.

The experimentally determined values for ρ for two different but related (1,3;1,4)-β-glucan preparations were –0.003 and 0.050. The data therefore indicated that the (1,3)-β-linked cellotriosyl and cellotetraosyl units were arranged essentially at random along the polysaccharide chain. p, the probability that a cellotriosyl unit is followed by another cellotriosyl unit. q, the probability that a cellotetraosyl unit is followed by a cellotriosyl unit.

The heterogeneity in fine structure imparted by this chemistry carries important consequences for the physicochemical and hence functional properties of the (1,3;1,4)-β-glucan. Thus, mid-level ratios of cellotriosyl/ cellotetraosyl units, say 1.5:2.5, would meet the functional requirements of a porous, gel-like wall matrix, whereas much higher or lower ratios would characterize a more conformationally regular, less soluble (1,3;1,4)-β-glucan that would have increased capacity to aggregate with other (1,3;1,4)-β-glucan molecules and possibly with cellulose and other wall polysaccharides such as heteroxylans.

The wide adoption of (1,3;1,4)-â-glucans in the *Poaceae* might be attributable to the ability of these species to synthesize (1,3;1,4)-β-glucans with mid-range ratios of cellotriosyl/cellotetraosyl units. Biological models for the synthesis of the (1,3;1,4)-β-glucans must account for their fine structural features, three of which challenge traditional cellular and biochemical views of how wall polysaccharides are synthesized.

These are:

i. The occurrence of single, isolated (1,3)-β-glucosyl residues amongst the predominating (1,4)-β-glucosyl residues,
ii. The random arrangement of cellotriosyl units and cellotetraosyl units along the chain
iii. The incorporation of longer blocks of up to 12-15 adjacent (1,4)-β-glucosyl residues in the molecule.

A two-phase assembly system has been proposed to account for the structural features of (1,3;1,4)-β-glucans. As illustrated diagrammatically in Figure, the first phase involves the assembly of a population of cellodextrins, which are linked to a gogli membrane–associated macromolecule. It is envisioned that a population of (1,4)-β-oligoglucosides is synthesized by the cellulose synthase–like CslF or CslH enzymes, some of which are located in the Golgi.

After transfer of the membrane-bound (1,4)-β-oligoglucosides to the plasma membrane, the second phase of the model suggests that the (1,4)-β-oligoglucosides are taken at random by another enzyme, possibly a callose synthase, a xyloglucan endotransglycosylase or even another CslF isoenzyme, and linked through (1,3)-β-linkages.

The model is consistent with the cellular location of the final (1,3;1,4)-β-glucan polysaccharide. During extensive immunochemical examination of (1,3;1,4)-β-glucan synthesis in a range of different cell types, the polysaccharide could not be detected in the Golgi, even where its concentration in the wall was very high.

The Golgi-located cellodextrins proposed to represent the first phase of (1,3;1,4)-β-glucan synthesis would not be detected by the (1,3;1,4)-β-glucan antibodies used. However, another group has recently disagreed and has concluded that some polymeric (1,3;1,4)-β-glucan is located in Golgi-derived vesicles of maize coleoptiles. We predict that these cell biological aspects of cell wall polysaccharide synthesis will remain the topic of debate for some time.

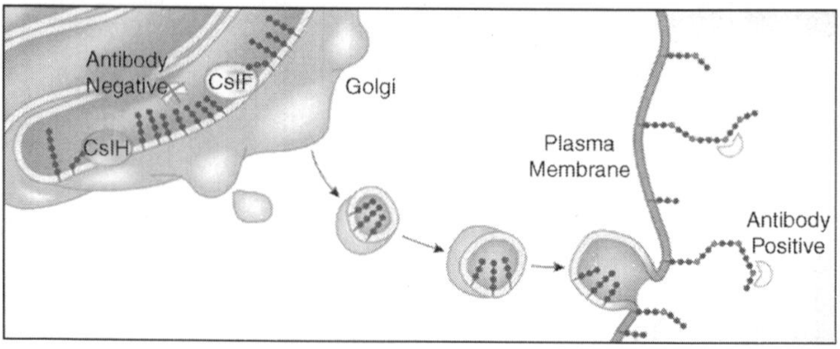

Fig. An Alternative, Two-phase Mechanism for Noncellulosic Polysaccharide Assembly.

Noncellulosic polysaccharides of the matrix phase in plant cell walls are generally considered to be synthesized in their entirety in the Golgi apparatus, from which they are transported via Golgi-derived vesicles to the plasma membrane. Fusion of the vesicles with the plasma membrane results in the deposition of the polysaccharides into the periplasmic space and their eventual incorporation into the wall. In an alternative theory, it has been proposed that oligosaccharides might be assembled in the Golgi, where they remain attached to a recyclable membrane lipid (or protein).

This would be consistent with early but not generally accepted indications that membrane-associated sitosterol or other sterols might act as intermediates in cellulose synthesis. The oligosaccharides would subsequently be polymerized by a plasma membrane–bound transferase that would deposit the nascent polysaccharide chain directly into the wall. Callose synthase and xyloglucan endotransglycosylases were suggested as candidate oligosaccharide polymerases.

In this diagram, a model for the potential assembly of (1,3;1,4)-β-glucans is presented, through which a population of different-length cellodextrins are synthesized in the Golgi by CslF and CslH enzymes. These cellodextrins, which the (1,3;1,4)-β-glucan antibody does not recognize, would then be transported to the plasma membrane via Golgi-derived vesicles. Upon fusion with the plasma membrane, the cellodextrins are released to allow the second stage os assembly to occur where they are joined by (1,3)-β-linkages. This yields the final polysaccharide now recognized by the (1,3;1,4)-β-glucan antibodies. It has been postulated that an assembly system similar to this might also apply to other noncellulosic polysaccharides of plant cell wall.

Although the theory of two-phase assembly of wall polysaccharides in the Golgi and at the plasma membrane was developed initially to account for experimental data related not only to the cell biology of (1,3;1,4)-β-glucan biosynthesis but also to the inbuilt heterogeneity in their fine structure, the model might be equally applicable to the synthesis of other wall polysaccharides. Xyloglucans have a strong repetitive element in their structures and might be considered a polymer of xylosylated tetraglucoside units. Pectic polysaccharides contain some repeating elements, and extracellular arabinogalactan proteins are believed to be made up of repeating oligosaccharide units.

Heterogeneity in Mannans

The cell wall polymer family for which fine structure has been most precisely defined and reconciled with biosynthesis processes is the galactomannans, members of which have a (1,4)-β-mannan backbone along which some but not all 6-C(OH) positions are substituted with á-galactose. Mannose/galactose ratios vary from about 1:1 to as high as 20:1.

On the basis of enzyme digestion and quantitative analysis of digestion-resistant oligosaccharides, it has been shown that second-order Markov chain statistics are sufficient to describe the fine structure of a range of galactomannans. For a processive biosynthesis mechanism in which galactosyl transferase activity is coupled with backbone mannan synthase, a second-order Markov chain is consistent with the galactomannan subsite recognition sequence of the galactosyltransferase encompassing at least three backbone mannosyl residues—that is, the site of substitution and the two preceding residues towards the reducing end of the growing galactomannan chain.

Datasets from three different plant species generated three distinctly different sets of probabilities and thus different galactosyl-substitution patterns. Subsequently, heterologously expressed galactosyl transferases in transgenic tobacco resulted in the production of galactomannan with Markov probability values characteristic of the original source of the galactosyl transferase. The Markov statistics describing polymer structure also allow a precise prediction of the probability of unsubstituted mannan segments of defined length, and these have been shown to be of consequence in the formation of interpolymeric junction zones that can form the basis for three-dimensional gel-like matrices of the type thought to be present within plant cell walls.

A Potential Role for Hydrolases in Creating Heterogeneity

An emerging theme in cell wall biology is the presence of polysaccharide hydrolases in tissues in which walls are being deposited and there is net synthesis of constituent wall polysaccharides. One example is the presence of (1,4)-β-glucan endohydrolases (cellulases) such as Korrigan in tissues in which cellulose synthesis is occurring.

The role of the cellulase has not been defined unequivocally, but it may be involved in the release of nascent cellulosic chains from the cellulose synthase enzyme complex, or it might be involved in the trimming of noncrystalline cellulose from the surface of microfibrils. The latter possibility is consistent with observations that cellulose crystallinity is decreased in

Arabidopsis lines that carry a silencing mutation in the *kor* gene. Molecular heterogeneity in heteroxylans might be generated by partial hydrolysis of the nascent polysaccharide rather than during the biosynthetic process itself. Newly synthesized arabinoxylans in maize and barley coleoptiles are usually highly substituted. Indeed, most if not all the xylosyl residues of the (1,4)-β-xylan backbone are initially substituted with arabinosyl residues. However, as the wall matures the degree of substitution decreases, and this has been attributed to the activity of arabinoxylan arabinofuranohydrolases, which can remove arabinosyl residues from the (1,4)-β-xylan backbone.

Irrefutable emerging evidence points to high levels of chemical, and hence structural, heterogeneity in the polysaccharide constituents of plant cell walls. The heterogeneity can be measured between walls in different parts of the same plant, between walls in a single tissue section and between regions within a single cell wall. In addressing the 'big questions' in plant cell wall biology in the immediate future, we will need to ensure that our explanations adequately rationalize heterogeneity as a crucial component of wall structure and function. Of the major unanswered questions facing cell wall biologists, the unequivocal identification of enzymes involved in the biosynthesis of the backbones of xyloglucans and heteroxylans is particularly pressing. Given the evidence for the participation of multienzyme complexes in cellulose biosynthesis, it will be important to investigate the highly likely possibility that the synthesis of noncellulosic wall polysaccharides also requires the coordinated action of several enzymes and other proteins. The latter question becomes even more intriguing in the case of substituted wall polysaccharides, in which enzymes involved in the synthesis of the backbone might form a complex with enzymes that simultaneously add monosaccharide or oligosaccharide substituents to the backbone. Might hydrolases also form a part of such complexes?

The recent observation that different isoforms of a single enzyme family can direct the synthesis of wall polysaccharides with different structures and thus different physicochemical properties raises the question of what the molecular mechanisms used by enzymes to synthesize wall polysaccharides in plants are. To thoroughly and accurately describe these mechanisms, three-dimensional structures of the synthases will be required, but at this stage there are no crystal structures of plant polysaccharide synthases of the GT2 group in the public databases.

The conventional view of the cellular steps involved in noncellulosic wall polysaccharide assembly has also been challenged, although proposed alternative pathways are supported largely by indirect data. It is likely that new technologies will enable us to investigate alternative models of wall polysaccharide assembly at the cellular level in the immediate future; the results of these investigations will be potentially transformational in terms of the way we view the cell biology of wall metabolism in plants. Similarly, emerging *in situ* methods will enable us to define in more detail the nature of polysaccharide-polysaccharide, polysaccharide-protein and polysaccharide-lignin interactions in the wall, to define whether these interactions involve covalent or noncovalent linkages and to identify the enzyme systems responsible for mediating the interactions. It has been known for some years that oligosaccharides released from wall polysaccharides by the action of microbial enzymes can act as signal molecules and, in particular, signal a loss of cell wall integrity that elicits a range of plant defence responses to pathogen attack. Given the role of the cell walls of both the microorganism and the plant as contact points during their first encounter, the question of cell wall integrity and oligosaccharide signaling will continue to be of importance in describing how plants resist biotic stress. The structural complexity and heterogeneity of pectins are consistent with a role for these polysaccharides in these processes, but are there other polysaccharides in the walls of the *Poaceae*, where pectin levels in walls are often very low, that assume a similar role?

Finally, it is important to acknowledge the recent appearance of two major research drivers that we can predict will be focused in the immediate future on the key questions outlined above. First, massive investments are being directed by both the private and public sectors towards the use of crop and plant residues, which consist in large part of cell wall material, as biomass sources for bioethanol and other biofuels production. Second, plant cell walls are attracting intense interest as a major source of dietary fibre in human foods, as increased fibre intake has been shown to significantly reduce the risks of such widespread and serious diet-related conditions as type 2 diabetes, cardiovascular disease, colorectal cancer and diverticular disease. We see a precise definition of heterogeneity and the underlying mechanisms for achieving the broad range of physicochemical properties as central for the ongoing biotechnological use of plant cell walls and their constituent polysaccharides.

APPLICATION OF PLANTS CELL EXPANSION

Within 5-10 minutes of application of auxins to isolated epicotyls, or intact tip of the plant or excised segments, growth in their length starts. This early effect is rapid and spontaneous. This rapid growth continues up to 30-45 minutes and then slows down but remains steady. This is also accompanied by an increase in cellular respiration, protein synthesis and the rapid extrusion of protons or in particular Hydrogen ions through the cell walls.

Another significant observation of this early rapid growth is insensitive to the inhibition of transcription (RNA synthesis) and translation (protein synthesis) - ex. Actinomycin-D at high concentration inhibits RNA synthesis completely and cycloheximide (CHI) inhibits cytosol protein synthesis. This clearly suggests that the early rapid growth requires neither the synthesis of RNA nor the synthesis of proteins. But this early effect of auxin is totally inhibited by colchicines at 0.5 to 1 per cent concentration. Colchicine, an alkaloid from the tubers of Colchicum autumnale brings about the inhibition of tubulin (TB) polymerisation into long microtubules, which act as the cytoskeleton for mechanical support in the cytoplasm. The colchicines effect on early growth further indicates the important role played by the assembly of microtubules in providing a mechanical structure and the force for all expansion.

Increased metabolic activity, particularly the degradation of osmotically inactive starch, etc., leads to the increase of osmotic pressure (they are related to each other) and a steep DPD gradient is created. This facilitates the inward diffusion of water; consequently turgour pressure builds up within the cells. This force acts on all sides of the cell wall and forces the cell to expand in length and breadth. However recent investigations have shown that the cell elongation takes place under negative pressure.

The above features suggest that the process or the mechanism of cell elongation is complex and multiphased, but highly regulated. Various theories have been proposed from time to time and none of them could prove their claim because of one or the other drawbacks. Here we will restrict to a comprehensive account of cell elongation.

Early phase: The auxin's effect in the early phase on cell growth is rapid. This is insensitive to the inhibitors of transcription and translation but insensitive to colchicines and respiratory inhibitors like cyanide, DNP etc. From this it can be deduced that cell elongation does not require new RNA and protein synthesis, but it requires ATP and ATP dependent tubulin polymerisation into microtubular cytoskeleton. Auxin at the early stages activates and increases the rate of respiratory metabolism.

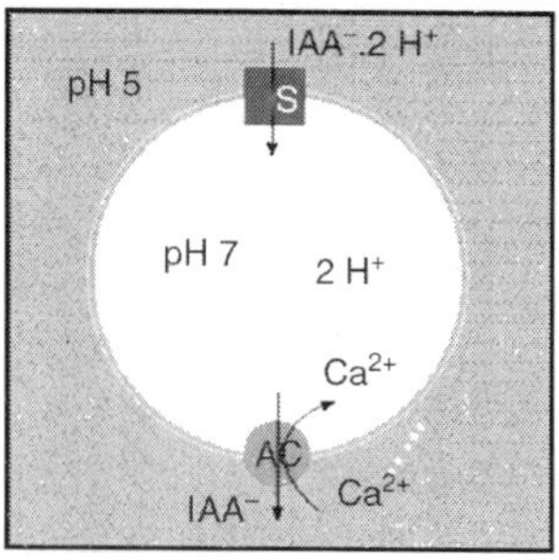

With the result ATP pool builds up rapidly. Auxin also activities membrane bound nucleating centre of tubulin polymerisation. These nucleating centres, utilising ATP and tubulin monomers found in the cytoplasm, start polymerising and soon build up long microtubules.

These, in turn, start orienting parallel to the long axis of the cell, which probably requires cytochalasin-B sensitive microfilaments. With the rapid increase in microtubules, oriented in longitudinal plane, considerable force is exerted on the cell boundaries.

As the microtubular polymerisation is building up, ATP ase/H* gets activated, and H* ions are pumped out into the interspaces of cellulose fibrils acidic pH in turn activates cellulases enzymes, which starts loosening up the cell-wall cellulose fibres. This loosening up of cell wall concomitant with acting as a mechanical force-the cell expands in length. This process starts 4-5 minutes after application of auxins and the elongation continues rapidly for about 20-30 minutes.

The late phase starts around 30-40 minutes after auxin application. This is slower than the first phase, but growth is steady. This phase is sensitive to Actinomycin-D, CHI, Colchicine and respiratory inhibitors, indicting that this process requires metabolic energy, tubulin-polymerisation, RNA synthesis as well as protein synthesis. As tubulin pool gets exhausted in the first phase, it has to be replaced. That means it requires the synthesis of RNA and proteins. Apart from these proteins, the cytoplasm requires the synthesis of other proteins, for the increase in the mass of protoplasm. It is quite possible that the early activation of early translation may have significant effect on gene activation. Thus it takes some time for the second phase of growth to set in. The second phase provides all materials for sustained growth.

Growth Movements: The phenomenon of bending movement of plants in response to light stimulus is called Phototropism, the growth of roots towards soil is called Geotropism.

The curvature growth due to the touch is named as Thigmotropism, etc. Such differential growth movements are due to differential concentration of auxin. The stem tips and root tips have different optima of their auxin requirement to show maximum growth. The mechanism of these movements will be explained in the chapter on growth movements. Auxin brings about phototropic movement because differential accumulation of auxin at exposed and darker side. *Root initiation*: Vegetative propagation of stem cuttings is a normal practice in horticulture, floriculture and agriculture. Out of some stem cuttings which are put into soil, a few produce roots easily but others may or not produce roots. The roots that develop from any part of the plant body other than the radicle are called adventitious roots.

The development of such roots from the stem or any other part is a phenomenon of dedifferentiation. Stem is quite different fro the root with respect to its structure and function. Still some of the tissues found in stem undergo dedifferentiation and develop into roots. This dedifferentiation and growth of roots from the stem is initiated by auxins Indole acetic acid, IBA, NAA, and other hormones. When applied to stem cuttings they induce the formation of roots.

IBA has been found to be very good in this regard.This indicates that the pre-existing m RNA protein complex is drawn towards protein synthesis: it shows increased rate of protein synthesis. Some of the proteins thus synthesised in turn activate differential nuclear gene activation. As a result of it, a cascade of events occurs. Generally there is an increase in the synthesis of all cellular proteins. At the same time there is an enhanced rate of synthesis of some specific proteins. Also some new proteins are synthesised. Among these, tubulin, a membrane bound fraction, is synthesised at a high rate. It is believed that this protein along with microfilaments gets assembled into microtubules which orient themselves in a specific way so as to facilitate the polarity fixation for cell division.

In response to auxins, some of the cells in the pericycle region get stimulated and the required macromolecules are synthesised by differential gene activation. Thus these cells endowed with these new potentialities divide and redivide, and soon they organise into root primordial. They in turn develop into new roots which penetrate through cortical cells. The time taken for this new root formation, in response to auxin supply is just 24-36 hours. This effect of auxin induced new root formation has been well exploited by horticulturists and others.

The following diagram shows how auxins interact with cytokinins in eliciting growth responses. *Apical dominance*: The active growth of terminal bud inhibits the development of lower axillary bunds. This effect is called apical dominance. Many plants like conifers and other exhibit the apical dominance so predominantly that the plant grows tall with few branches and appear conical in shape. But plants like banyan, mango, tamarind do not show that much of apical dominance, hence these plants spread outwardly. This phenomenon of apical dominance has been attributed to the differential concentration of auxin and nutrient supply to the axillary buds.

Abscission: With the onset of winter season the leaves of deciduous plants turn yellow and fall off. The falling of the leaves is due to the formation of abscission layers at the base of the petioles. This layer consists of thin walled but suberised cells. The middle wall gets disintegrated and the leaves fall down. The development of this layer is stimulated by a plant hormone called ABA. But the application of auxin to such leaves prevent the formation of abscission layer and thus prevent the falling off the leaves. This has a greater application particularly in preventing the premature falling of fruits like apple, orange, etc.

Parthenocarpy: Development of fruits without fertilization is referred to as parthenocarpy and such fruits are seedless. Normally plants do not produce parthenocarpic fruits. Normally plants do not produce parthenocarpic fruits. But the application of auxin to flowers induces parthenocarpic fruit setting and prevents premature falling of fruits as well. Plants with triploid and many polyploid genomes do not set fruits. If such plants are sprayed with auxins at the time of flowering, fruit setting is enhanced. The fruits thus developed are large, seedless, and sweeter. Seedless oranges, lime, watermelon, sapota, ramphal and seethaphal, if they are produced this way, they have good market at the commercial level.

Gibberllic Acid

Gibberllic Acid is another natural hormone found distributed in higher plants. This was first discovered in Japan, later its structural and functional details have been elucidated by the people in the west. So far 52 or more different kinds have been identified. Among them GA2 2GA1 are the most common ones found in all plants. GA is a steroid like molecule in its structure. Young leaves are the sites in which this hormone is mediated by phytochromes and then it is transported to the regions of growth.

The most significant effect of GA on the growth of plants is it effects internodal cell elongation profoundly than any other hormones. Even genetic dwarfs (*e.g.* Dwarf Pisum sativum) grow as tall as the natural tall plants. It also brings about a dramatic effect called Bolting and Flowering. Long day plants (which require 12-14 hrs of day light treatment for flowering) under short day conditions produce a cluster of leaves on highly shortened stem and never produce flowers. When such plants are sprayed with GA, the stem starts elongation with enhanced growth of internodes. Simultaneously the terminal buds starts producing floral branches. This phenomenon is called bolting and flowering. However both bolting and flowering are separated by time and space. Besides these above said effects, GA is essential for the mobilisation of food reserves for the germinating cereal grains like wheat and maize. At the time of germination, the young embryonic axis synthesises this hormone, which is soon released into endosperm and reaches the aleurone layer.

It is this layer, GA activates differential gene expression. As a result, specific m.RNAs is transcribed and later specific proteins like amylase are synthesised on translation. The amylase thus synthesised, acts upon the starch found in the endosperm, and the glucose thus released is drawn into the embryonic tissue which is in dire need of energy source. Notwithstanding the above effects, GA also effects parthenocarpy, rooting, germination, etc. This hormone also interacts with other natural hormones like auxin and cytokinin and brings about differential growth.

CYTOKININS

Another class of growth regulating hormones has been discovered by Miller et al and they are called Cytokinins. They are the derivatives or nucleotides. 6-Furfuryl amino purines, Benzyl amino purines and such compounds act as cytokinins.

Cytokinins are generally synthesised in the root tips, and then they are translocated to other regions. The endosperm of coconut milk, Borassus and of other plants is a rich source of cytokinins. Cytokinin level in plants is further regulated by other hormones like Auxins and GAs. On the contrary cytokinins control the synthesis of other hormones like Abscisic acid, a growth inhibitor.

Cytokinin *per se* has an important role in the cytokinesis. But along with auxins, it effects differentiation of plant organs from the callus. Cytokinin by itself inhibits root formation, but in the presence of auxins, it induces organ formation like roots and shoots. If the ratio between auxin and cytokinin is high, roots are differentiated, but the lower ratio induces the formation of shoots.

The intermediate ratio stimulates the development of these hormones in differentiation at the molecular level is very interesting by unfortunately it is least understood. Cytokinins are also found to break dormancy of winter buds as well seeds. The exact mechanism by means of which they over come the dormancy is not known, still it has been thought, that they effect through the suppression of IBA which is responsible for dormancy. Another interesting effect that the cytokinins produce is the prolongation of senescence (Ageing). This effect is often referred to as Richmond and Lang's effect.

APPLICATION OF IMMOBILIZED CELLS

Immobilization of plant cells is considered to be of importance in research and development in plant cell cultures, because of the potential benefits that could be provided:

- The extended viability of cells in the stationary (and producing) stage, enabling maintenance of biomass over a prolonged time period;
- Simplified downstream processing (if products are secreted);
- The (putative) promotion of differentiation, linked with enhanced secondary metabolism;
- Higher cell density enabling a reduced bioreactor size, thereby reducing costs and the risk of contamination;
- Reduced shear sensitivity (especially with entrapped cells);
- Promotion of secondary metabolite secretion, in some cases;
- Flow-through reactors can be used enabling greater flow rates;
- Minimization of fluid viscosity increase, which in cell suspension causes mixing and aeration problems.

An immobilization system which could maintain viable cells over an extended period of time and release the bulk of the product into the extracellular medium in a stable form, could dramatically reduce the costs of phytochemicals production in plant cell culture. However, an immobilized system also has the problems described below:

- Immobilization is normally limited to cases where production is decoupled from cell growth;
- The initial biomass must be grown in suspension;
- Secretion of product into the extracellularly medium is imperative;
- Where secretion occurs there may be problems of extracellular degradation of the products;

- When gel entrapment is used, the gel matrix introduces an additional diffusion barrier.

Due to these problems, a system with commercial potential has not yet been developed in plant tissue cultures. However, various immobilization methods have been developed, ie., entrapment, adsorption and covalent coupling.

Some preliminary results have been obtained with immobilized cells. Early work with C. roseus, showed that agar, agarose and carageenan were all suitable immobilization matrices suitable for maintenance of cell viability; but alginate was superior in terms of ajmalicine production. The accumulation of serpentine by C. roseus and anthraquinones by Morinda citrifolia were both enhanced in the immobilized state when compared with freely suspended cells.

It should be noted however, that the possibility that alginate acts as an elicitor of secondary metabolism cannot be ruled out. Agar has been shown to stimulate shikonin accumulation in L. erythrorhizon cultures. Lambe and Rosevear have successfully immobilized C. roseus cells in polyacrylamide with alginate and observed prolonged viability and increased productivity. Adsorption immobilization has been successfully used with a number of plant species. Capsicum frutescens cells immobilized on polyurethane foam produced 50 times as much capsaicin as suspension cells. Similarly, Solanum nigrum cells accumulated glycoalkaloids to levels exceeding those found in suspensions. Datura innoxia cells accumulated tropane alkaloids with a profile similar to that of the intact plant, whilst in free suspensions productivity was markedly suppressed. In general it appears that mild immobilization either through gel entrapment or surface adsorption enhances productivity and prolongs the viability of cultured cells.

As described, immobilized cells can also be used as biocatalysts for biotransformations. Such a system compares favourably with the use of freely suspended cells since, in the case of immobilization, the catalyst is theoretically reusable and the product is easily separated from the biomass. The most appropriate example is that of the 12-hydroxylation of ß-methyldigitoxin to ß-methyldigoxin with alginate-entrapped Digitalis lanata cells. The enzyme activity was maintained by the immobilized cultures for a period of 61 days. Furthermore, the product was located in the extracellular medium. Mild permeabilization of the cells may enable biotransformation rates to be increased.

Polyurethane-immobilized C. frutescens cells fed capsaicin precursors produced this metabolite at levels of up to 10 times those of non-fed cultures. DiCosmo *et al*. found that glass fibres can be used as a carrier of plant cells to produce useful plant metabolites. Papaver somniforum cells were immobilized on fabric of loosely woven polyester fibres arranged in a spiral configuration on stainless steel support frame by Kurz *et al*. to produce sanguinarine, an antibiotic in oral hygiene. The yield was 3.6 mg/g-fw. by immobilized cells and was more than twice as much as by suspension cells.

PRODUCT SECRETION

Many plant products produced by cell cultures have been reported to be accumulated intracellularly. However, it may be possible to produce much higher level of product if it was secreted into the medium. This is because the product intracellularly accumulated sometimes inhibits its own synthesis by regulation mechanisms such as product inhibition and repression. For many immobilized plant cell systems to work it is essential that a significant amount of product is released into the medium. Two types of immobilization system with C. roseus *i.e.* gel entrapment in polysaccharide beads and in polyacrylamide sheets, have both exhibited alkaloid release by mechanisms that do not appear to be associated with losses in viability. Capsicum frutescens cells immobilized on polyurethane released capsaicin entirely into the medium, although other species immobilized by the same method retained the product intracellularly.

To enhance release, permeabilization of cell membranes has been attempted, but with only limited success. Brodelius tested five permeabilizing agents on three different species, and although product release was achieved, cell viability dropped in most cases. The exceptions were DMSO and Triton X-100, applied to C. roseus cells. Other attempts to permealize cells have also resulted in non-viable populations and the use of electroporation has also resulted in a viability decrease. The release of betanin by Beta vulgaris cells ultrasonicated for 20 to 60 seconds has been reported, with no apparent effect on cell viability. Chenopodium rubrum cells, immobilized in alginate beads, secreted the red betacyanin pigment amaranthin into the medium. However, the pigment was subsequently degraded; chitosan and DMSO permitted further product release into the extracellular medium, but this was also accompanied by product degradation. Low concentrations of chitosan (0.01%) and DMSO (5.7%) incubated with the cultures for 96 hr did not appear to affect viability significantly, but a longer incubation period had a deterlerious effect.

The lack of an appropriate means of product release is a serious problem in terms of an industrial approach particularly based on immobilization of plant cells. It is perhaps necessary to examine in vivo physiological mechanisms of release. Thus, some researchers have been investigating the factors involved in vacuolar phytochemical storage. The accumulation of indole alkaloids in C. roseus vacuoles has been attributed to an ion-trap mechanism whereby the basic indole alkaloids are trapped in the acidic vacuole due to their positive charge at low pH, preventing diffusion across the tonoplast. This mechanism has also been demonstrated for quinoline alkaloid accumulation in Cinchona species.

Active uptake mechanisms have also been reported for indole alkaloids in C. roseus vacuoles and isoquinoline alkaloids in Fumaria capreolata L. The physiological significance of these two possible mechanisms is not yet clear, since there is obviously strong evidence for both. In terms of product release, it is pertinent to note that in cell cultures, an efflux of alkaloids was observed under certain conditions, indicating an equilibrium between the intracellular and extracellular compartments that could be perturbed by medium acidification with subsequent product release. The release of serpentine by C. roseus cells was observed when the cells were filtered and resuspended in fresh or conditioned medium and it was suggested that temporary membrane uncoupling was responsible. Although most plant cell culture products were shown to be accumulated intracellularly, Shuler *et al.* recently reported that almost all of taxol produced by Taxus brevifolia cell cultures was detected in the culture filtrate. This is an exceptional case in plant tissue culture research.

MUTAGENESIS

In the fermentation industry, induction of genetic mutant strains of microorganisms is ubiquitous, and auxotrophic and/or regulatory mutants are used extensively to produce a variety of products including amino acids, nucleotides, antibiotics, etc. However, mutagenesis has limited applicability to plant cell cultures, because of their diploid genetic make-up: the chance of obtaining a double mutation in a target gene is less than 1 in 106. Although, in principle, haploid plants can be produced from anther cultures; in practice haploid cell cultures tend to revert to the diploid state. This makes the chance of isolating over-producing cells from mutagen treatment of haploid cells very low. Furthermore, biosynthetic pathways of many secondary metabolites and their regulation mechanisms in higher plants are not always understood precisely, therefore, it is also difficult to know what kind of mutants should be induced in order to increase product synthesis.

However, Berlin induced p-fluorophenylalanine resistant cell lines of tobacco cell cultures and found that five lines of N. tabacum and five lines of N. glauca out of 31 resistant cell lines accumulated higher levels of phenolics. He and his colleagues also reported that a resistant strain of N. tabacum produced 6 to 10 times higher levels of cinnamoyl putrescine than that of the parent strain.

Figure shows the comparison of enzyme activities involved in the biosynthesis of cinnamoyl putrescines in a parent strain, TX1 and a resistant strain, TX4. Enzyme activities in the p-fluorophenylalanine resistant strain were much higher than those of the parent. Berlin *et al.* also found that a parent strain of C. roseus produced catharanthine only in the production medium but its tryptophan analog resistant mutant accumulated the same alkaloid even in the growth medium. Among several types of analogues tested, 4-fluorotryptophan was found to be the most efficient of them.

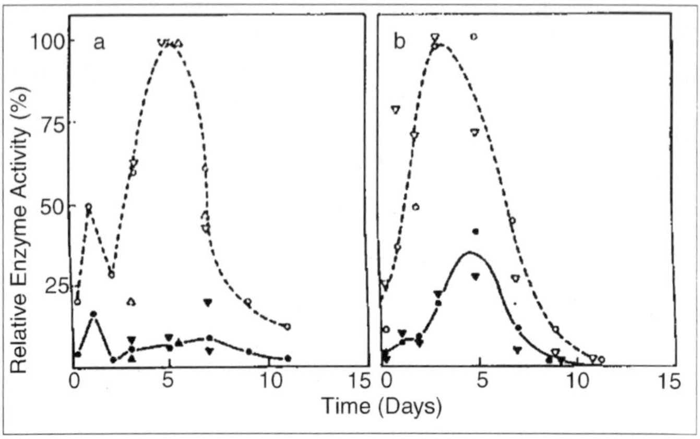

Fig. Comparison of Enzyme Activities

Increase of metabolite levels using regulatory mutants is theoretically possible and selection of suitable analogues for this purpose could be an important factor in order to produce a variety of products. Several research groups have used mutagens such as N-methyl-N'-nitro-N-nitrosoguanidine, ethylmethane sulfonate, N-ethyl-N-nitrosourea or X-rays to induce high yielding strains.

In China, a fine callus strain of Anisodus acutangulus was derived through irradiation with 4000 R of X-ray, and the level of scopolamine in the cells was 0.177 mg/g d.wt., which was about 30% higher than of the parent. The productivity was stable. Deus treated C. roseus with X-rays and obtained a cell line having a high producing activity of serpentine whose level was 2%.

A cell line of Lavendula vera producing a high level of free biotin was obtained by gamma ray irradiation for 1 hr. with 60_{Co} (dose 10 KR h) by Watanabe *et al*. The line was found to contain 7 times the amount, 0.425 μg/g-fresh wt., of free biotin compared with the original unselected cells, 0.061 μg/g fresh wt., and 4.5 times that found in the leaves. In spite of the difficulty of obtaining auxotrophic mutants in plant cells, several groups have been actively working with this approach using haploid cells or protoplasts although their aim is not always for plant metabolite production. Induction of a mutant having altered permeability could also be important, because plant cells generally accumulate their metabolites intracellularly, which is disadvantageous in commercial production because the amount of compounds produced is usually low.

If the cells excreted products in large amounts, the product cost would be reduced. According to Berlin's results, Thuja occidentalis excreted monoterpenoids, but the levels in the medium were only 5% of those in the mother plant. The same group reported that Macleya microcarpa cells excreted nearly all the alkaloids detectable in the culture flask. Yamakawa *et al.*

cultivated Tinospora rumphii cells in 14 ml medium and found 0.57 (5.3% in dry weight cells) of isoquinoline alkaloids in the cells and 0.5 mg in the culture filtrate after 7 days cultivation. By replacement of the medium with fresh medium after 3 days of cultivation, 0.50 mg of the alkaloids in the cells, and 1.02 mg in the filtrate were accumulated. It seems then that the products must be excreted to some extent. Unfortunately, the secretion mechanisms for secondary products in higher plant cells have not yet been elucidated and extensive fundamental studies are required before meaningful manipulation can take place.

CELL BIOLOGY DEFINITION OF CELL MEMBRANE

The cell membrane is a thin semi-permeable membrane that surrounds the cytoplasm of a cell, enclosing its contents. Its function is to protect the integrity of the interior of the cell by allowing certain substances into the cell, while keeping other substances out. It also serves as a base of attachment for the cytoskeleton in some organisms and the cell wall in others. Thus the cell membrane also serves to help support the cell and help maintain its shape.

The cell membrane is primarily composed of a mix of proteins and lipids. While lipids help to give membranes their flexibility, proteins monitor and maintain the cell's chemical climate and assist in the transfer of molecules across the membrane. Phospholipids are a major component of cell membranes. They form a lipid bilayer in which their hydrophillic (attracted to water) head areas spontaneously arrange to face the aqueous cytosol and the extracellular fluid, while their hydrophobic (repelled by water) tail areas face away from the cytosol and extracellular fluid. The lipid bilayer is semi-permeable, allowing only certain molecules to diffuse across the membrane.

CYTOPLASM

The cytoplasm consists of all of the contents outside of the nucleus and enclosed within the cell membrane of a cell. This includes the cytosol and in euckaryotic cells, organelles such as mitochondria and ribosomes. Also located within the cytoplasm is the cytoskeleton, a network of fibres that help the cell maintain its shape and give it support. The cytoplasm is clear in colour and has a gel-like appearance. It is composed mainly of water and also contains enzymes, salts, organelles, and various organic molecules. The cytoplasm helps to move materials around the cell and also dissolves cellular waste.

THE CYTOSKELETON

In Journey into the Cell, we looked at the structure of the two major types of cells: prokaryotic and eukaryotic cells. Now we turn our attention to the "infrastructure" of a eukaryotic cell, the cytoskeleton.

What is the Cytoskeleton?

The cytoskeleton is a network of fibres throughout the cell's cytoplasm that helps the cell maintain its shape and gives support to the cell. A variety of cellular organelles are held in place by the cytoskeleton.

Cytoskeleton: Distinguishing Characteristics

The cytoskeleton is composed of at least three different types of fibres: microtubules, microfilaments and intermediate filaments. These types are distinguished by their size with microtubules being the thickest and microfilaments being the thinnest.

- Microtubules are hollow rods functioning primarily to help support and shape the cell and as "routes" along which organelles can move. Microtubules are typically found in all eukaryotic cells.
- Microfilaments or actin filaments are solid rods and are active in muscle contraction. Microfilaments are particularly prevalent in muscle cells but similar to microtubules, they are also typically found in all eukaryotic cells.
- Intermediate filaments can be abundant in many cells and provide support for microfilaments and microtubules by holding them in place.

In addition to providing support for the cell, the cytoskeleton is also involved in cellular motility and in moving vesicles within a cell, as well as assisting in the formation of food vacuoles in the cell.

ENDOPLASMIC RETICULUM

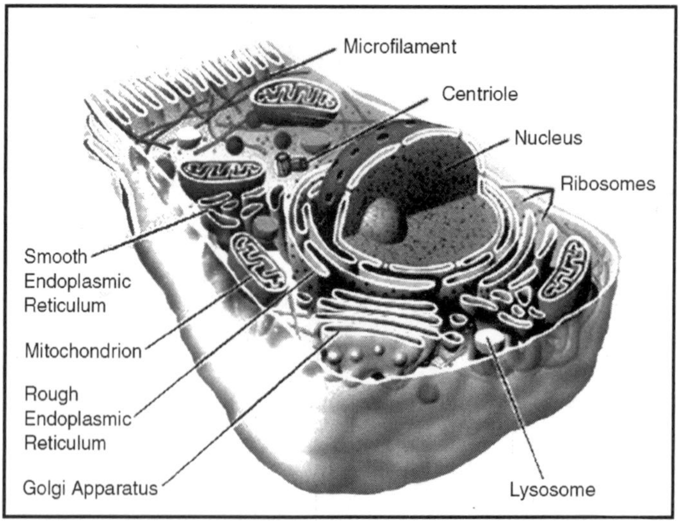

Fig. Panagene.comCustom PNA Synthesis, Fmoc Monomer, Telomere FISH, MiRNA Inhibitors

Endoplasmic Reticulum

In Journey into the Cell, we looked at the structure of the two major types of cells: prokaryotic and eukaryotic cells. Now we turn our attention to the endoplasmic reticulum (ER). The endoplasmic reticulum is a network of tubules and flattened sacs that serve a variety of functions in the cell. There are two regions of the ER that differ in both structure and function. One region is called rough ER because it has ribosomes attached to the cytoplasmic side of the membrane.

The other region is called smooth ER because it lacks attached ribosomes. Typically, the smooth ER is a tubule network and the rough ER is a series of flattened sacs. The space inside of the ER is called the lumen. The ER is very extensive extending from the cell membrane through the cytoplasm and forming a continuous connection with the nuclear envelope. Since the ER is connected with the nuclear envelope, the lumen of the ER and the space inside the nuclear envelope are part of the same compartment.

Rough Endoplasmic Reticulum

The rough endoplasmic reticulum manufactures membranes and secretory proteins. In certainleukocytes (white blood cells), the rough ER produces antibodies. In pancreatic cells, the rough ER produces insulin. The rough and smooth ER are usually interconnected and the proteins and membranes made by the rough ER move into the smooth ER to be transferred to other locations.

Smooth Endoplasmic Reticulum

The smooth ER has a wide range of functions including carbohydrate and lipid synthesis. It serves as a transitional area for vesicles that transport ER products to various destinations. In liver cells the smooth ER produces enzymes that help to detoxify certain compounds. Inmuscles the smooth ER assists in the contraction of muscle cells, and in brain cells it synthesizes male and female hormones.

Eukaryotic Cell Structures

The endoplasmic reticulum is only one component of a cell.

The following cell structures can also be found in a typical animal eukaryotic cell:

- Centrioles - help to organise the assembly of microtubules.
- Chromosomes - house cellular DNA.
- Cilia and Flagella - aid in cellular locomotion.
- Cell Membrane - protects the integrity of the interior of the cell.
- Golgi Complex - manufactures, stores and ships certain cellular products.
- Lysosomes - digest cellular macromolecules.
- Mitochondria - provide energy for the cell.
- Nucleus - controls cell growth and reproduction.

- Peroxisomes - detoxify alcohol, form bile acid, and use oxygen to break down fats.
- Ribosomes - responsible for protein production via translation.

GOLGI COMPLEX

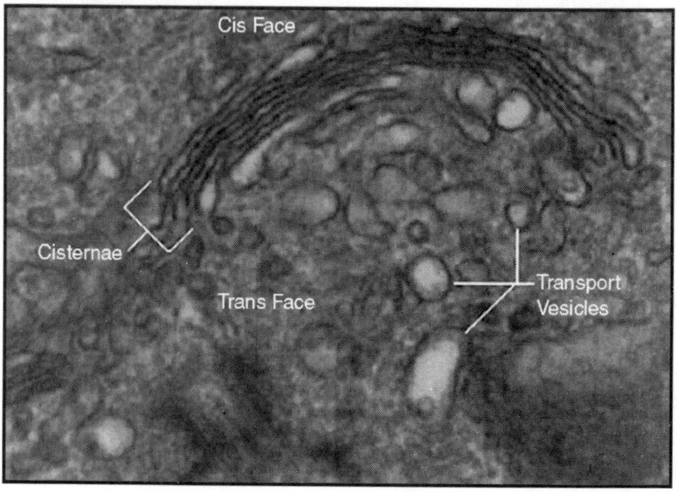

Fig. The Golgi Complex is Composed of Flat Sacs Known as Cisternae. The Sacs are Stacked in a Bent, Semicircular Shape.

In Journey into the Cell, we looked at the structure of the two major types of cells: prokaryotic and eukaryotic cells. Now we turn our attention to the "manufacturing and shipping centre" of a eukaryotic cell, the Golgi complex. The Golgi complex or Golgi apparatus is responsible for manufacturing, warehousing, and shipping certain cellular products, particularly those from the endoplasmic reticulum (ER).

Depending on the type of cell, there can be just a few complexes or there can be hundreds. Cells that specialise in secreting various substances typically have a high number of Golgi complexes.

Golgi Complex: Distinguishing Characteristics

A Golgi complex is composed of flat sacs known as cisternae. The sacs are stacked in a bent, semicircular shape. Each stacked grouping has a membrane that separates its insides from the cell's cytoplasm. Golgi membrane protein interactions are responsible for its unique shape. These interactions generate the force that shapes this organelle.

The Golgi complex is very polar. Membranes at one end of the stack differ in both composition and in thickness from those at the other end. One end (cis face) acts as the "receiving" department while the other (trans face) acts as the "shipping" department. The cis face is closely associated with the ER.

Golgi Complex: Molecule Transport and Modification

Molecules synthesized in the ER exit via special transport vesicles which carry their contents to the Golgi complex. The vesicles fuse with Golgi cisternae releasing their contents into the internal portion of membrane. The molecules are modified as they are transported between cisternae layers. It is thought that individual sacs are not directly connected, thus the molecules move between cisternae through a sequence of budding, vesicle formation, and fusion with the next Golgi sac.

Once the molecules reach the trans face of the Golgi, vesicles are formed to "ship" materials to other sites. The Golgi complex modifies many products from the ER including proteins and phospholipids. The complex also manufactures certain biological polymers of its own. The Golgi complex contains processing enzymes which alter molecules by adding or removing carbohydratesubunits. Once modifications have been made and molecules have been sorted, they are secreted from the Golgi via transport vesicles to their intended destinations. Some of the molecules are destined for the cell membrane where they aid in membrane repair and intercellular signalling. Other molecules are secreted to areas outside of the cell.

Transport vesicles carrying these molecules fuse with the cell membrane releasing the molecules to the exterior of the cell. Still other vesicles contain enzymes that digest cellular components. These vesicle form cell structures called lysosomes. Molecules dispatched from the Golgi may also be reprocessed by the Golgi.

Golgi Complex Assembly

The Golgi complex is capable of disassembly and reassembly. During the early stages of mitosis, the Golgi disassembles into fragments which further breakdown into vesicles. As the cell progresses through the division process, the Golgi vesicles are distributed between the two forming daughter cells by spindle microtubules. The Golgi complex reassembles in the telophase stage of mitosis. The mechanisms by which the Golgi complex assembles are not yet understood.

Cell Structures

The Golgi complex is only one component of a cell.

The following cell structures can also be found in a typical animal cell:

- Cell Membrane - protects the integrity of the interior of the cell.
- Centrioles - help to organise the assembly of microtubules.
- Chromosomes - house cellular DNA.
- Cilia and Flagella - aid in cellular locomotion.
- Endoplasmic Reticulum - synthesizes carbohydrates and lipids.
- Lysosomes - digest cellular macromolecules.
- Mitochondria - provide energy for the cell.
- Nucleus - controls cell growth and reproduction.

- Peroxisomes - detoxify alcohol, form bile acid, and use oxygen to break down fats.
- Ribosomes - responsible for protein production via translation.

MICROTUBULES

Microtubules are fibrous, hollow rods, that function primarily to help support and shape the cell. They also function as routes along which organelles can move. They are typically found in all eukaryotic cells and are a component of the cytoskeleton, as well as cilia and flagella.

Examples

Microtubules play a huge role in movement within a cell. They form the spindle fibres that manipulate and separate chromosomes during mitosis. Examples of microtubule fibres that assist in cell division include polar fibres and kinetochore fibres.

MITOCHONDRIA

Mitochondria

In Journey into the Cell, we looked at the structure of the two major types of cells: prokaryotic and eukaryotic cells. Now we turn our attention to the "power houses" of a eukaryotic cell, the mitochondria. Mitochondria are the cell's power producers. They convert energy into forms that are usable by the cell. Located in the cytoplasm, they are the sites of cellular respiration which ultimately generates fuel for the cell's activities. Mitochondria are also involved in other cell processes such as cell division and growth, as well as cell death.

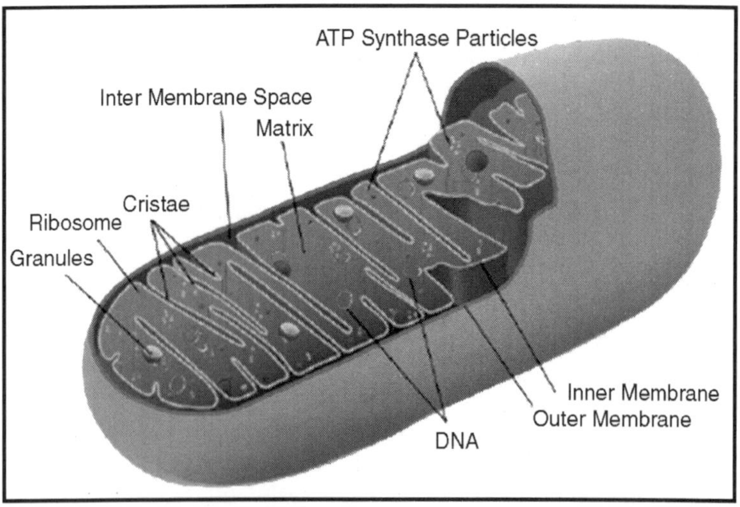

Fig. Mitochondrion

Mitochondria: Distinguishing Characteristics

Mitochondria are bounded by a double membrane. Each of these membranes is a phospholipidbilayer with embedded proteins. The outermost membrane is smooth while the inner membrane has many folds. These folds are called cristae. The folds enhance the "productivity" of cellular respiration by increasing the available surface area.

The double membranes divide the mitochondrion into two distinct parts: the intermembrane space and the mitochondrial matrix. The intermembrane space is the narrow part between the two membranes while the mitochondrial matrix is the part enclosed by the innermost membrane. Several of the steps in cellular respiration occur in the matrix due to its high concentration of enzymes.

Mitochondria are semi-autonomous in that they are only partially dependent on the cell to replicate and grow. They have their own DNA, ribosomes and can make their own proteins. Similar to bacteria, mitochondria have circular DNA and replicate by a reproductive process called fission.

BIOTECHNOLOGY IN PLANT CELL

Clonal propagation refers to the process of asexual reproduction by multiplication of genetically identical copies of individual plants. The vegetative propagation of plants is labour-intensive, low in productivity and seasonal. The tissue culture methods of plant propagation, known as 'micropropagation' utilizes the culture of apical shoots, axillary buds and meristems on suitable nutrient medium.The regeneration of plantlets in cultured tissue was described by Murashige in 1974. Fossard gave a detailed account of stages of micropropagation.

The micropropagation is rapid and has been adopted for commercialization of important plants such as banana, apple, pears, strawberry, cardamom, many ornamentals (*e.g.* Orchids) and other plants.The micropropagation techniques are preferred over the conventional asexual propagation methods because of the following reasons: (a) In the micropropagation method, only a small amount of tissue is required to regenerate millions of clonal plants in a year., (b) micropropagation is also used as a method to develop resistance in many species., (c) in vitro stock can be quickly proliferated as it is season independent,. (d) long term storage of valuable germplasm possible.

The steps in micropropagation method are:

- *Initiation of culture*: From an explant like shoot tip on a suitable nutrient medium,
- Multiple shoots formation from the cultured explant,
- Rrooting of in vitro developed shoots and,
- Transplantation-transplantation to the field following acclimatization.

The factors that affect micropropagation are:

- Genotype and the physiological status of the plant *e.g.* plants with vigorous germination are more suitable for micropropagation.,
- The culture medium and the culture environment like light, temperature etc. For example an illumination of 16 hours a day and 8 hours night is satisfactory for shoot proliferation and a temperature of 250C is optimal for the growth.

The benefits of micropropagation this method are:

- Rapid multiplication of superior clones can be carried out through out the year, irrespective of seasonal variations.
- Multiplication of disease free plants *e.g.* virus free plants of sweet potato (Ipomea batatus), cassava (Manihot esculenta)
- Multiplication of sexually derived sterile hybrids
- It is a cost effective process as it requires minimum growing space.

SOMACLONAL VARIATION

The genetic variations found in the in vitro cultured cells are collectively referred to as somaclonal variation and the plants derived from such cells are called as 'somaclones'. It has been observed that the long-term callus and cell suspension culture and plants regenerated from such cultures are often associated with chromosomal variations. It is this property of cultured cells that finds potential application in the crop improvement and in the production of mutants and variants (*e.g.* disease resistance in potato).

Larkin and Scowcroft working at the division of Plant Industry, C.S.I.R.O., Australia gave the term 'somaclones' for plant variants obtained from tissue cultures of somatic tissues. Similarly, if the tissue from which the variants have been obtained is having gametophytic origin such as pollen or egg cell, it is known as 'gametoclonal' variation.

They explained that it may be due to:

- Reflection of heterogeneity between the cells and explant tissue,
- A simple representation of spontaneous mutation rate, and
- Activation by culture environment of transposition of genetic materials.

Shepard *et al.* also contributed by screening about 100 somaclones produced from leaf protoplasts of Russet Burbank. They found that there was a significant amount of stable variation in compactness of growth habit, maturity, date, tuber uniformity, tuber skin colour and photoperiodic requirements.

Somaclonal Variations has been used in plant breeding programmes where the genetic variations with desired or improved characters are introduced into the plants and new varieties are created that can exhibit disease resistance, improved quality and yield in plants like cereals, legumes, oil seeds tuber crops etc. Somaclonal variation is applicable for seed

APPLICATIONS OF SOMACLONAL VARIATIONS

- Methodology of introducing somaclonal variations is simpler and easier as compared to recombinant DNA technology.
- Development and production of plants with disease resistance *e.g.* rice, wheat, apple, tomato etc.
- Develop biochemical mutants with abiotic stress resistance *e.g.* aluminium tolerance in carrot, salt tolerance in tobacco and maize.
- Development of somaclonal variants with herbicide resistance *e.g.* tobacco resistant to sulfonylurea
- Development of seeds with improved quality *e.g.* a new variety of Lathyrus sativa seeds (Lathyrus Bio L 212) with low content of neurotoxin.
- Bio-13 – A somaclonal variant of Citronella java (with 37% more oil and 39% more citronellon), a medicinal plant has been released as Bio-13 for commercial cultivation by Central Institute for Medicinal and Aromatic Plants.
- Supertomatoes- Heinz Co. and DNA plant Technology Laboratories (USA) developed Supertomatoes with high solid component by screening somaclones which helped in reducing the shipping and processing costs.

Production of Virus Free Plants

The viral diseases in plants transfer easily and lower the quality and yield of the plants. It is very difficult to treat and cure the virus infected plants therefore te plant breeders are always interested in developing and growing virus free plants.

In some crops like ornamental plants, it has become possible to produce virus free plants through tissue culture at the commercial level.

This is done by regenerating plants from cultured tissues derived from:

- Virus free plants,
- Meristems which are generally free of infection - In the elimination of the virus, the size of the meristem used in cultures play a very critical role because most of the viruses exist by establishing a gradient in plant tissues. The regeneration of virus-free plants through cultures is inversely proportional to the size of the meristem used.,
- Meristems treated with heat shock (34-36°C) to inactivate the virus,
- Callus, which is usually virus free like meristems.
- Chemical treatment of the media- attempts have been made to eradicate the viruses from infected plants by treating the culture medium with chemicals *e.g.* addition of cytokinins suppressed the multiplication of certain viruses.

Among the culture techniques, meristem-tip culture is the most reliable method for virus and other pathogen elimination. Viruses have been eliminated from a number of economically important plant species which has resulted in a significant increase in the yield and production *e.g.* potato virus X from potato, mosaic virus from cassava etc. These virus free plants are not disease resistant so there is a need to maintain stock plants to multiply virus free plants whenever required.

Production of Synthetic Seeds

In synthetic seeds, the somatic embryos are encapsulated in a suitable matrix (*e.g.* sodium alginate), along with substances like mycorrhizae, insecticides, fungicides and herbicides. These artificial seeds can be utilized for the rapid and mass propagation of desired plant species as well as hybrid varieties.

The major benefits of synthetic seeds are:

- They can be stored up to a year with out loss of viability
- Easy to handle and useful as units of delivery
- Can be directly sown in the soil like natural seeds and do not need acclimatization in green house.

Mutant Selection

An important use of cell cultures is in mutant selection in relation to crop improvement. The frequency of mutations can be increased several fold through mutagenic treatments and millions of cells can be screened. A large number of reports are available where mutants have been selected at cellular level. The cells are often selected directly by adding the toxic substance against which resistance is sought in the mutant cells. Using this method, cell lines resistant to amino acid analogues, antibiotics, herbicides, fungal toxins etc have actually been isolated.

Production of Secondary Metabolites

The most important chemicals produced using cell culture are secondary metabolites, which are defined as' those cell constituents which are not essential for survival'. These secondary metabolites include alkaloids, glycosides (steroids and phenolics), terpenoids, latex, tannins etc. It has been observed that as the cells undergo morphological differentiation and maturation during plant growth, some of the cells specialize to produce secondary metabolites. The in vitro production of secondary metabolites is much higher from differentiated tissues when compared to non-differentiated tissues. The cell cultures contribute in several ways to the production of natural products.

These are:

- A new route of synthesis to establish products *e.g.* codeine, quinine, pyrethroids,

- A route of synthesis to a novel product from plants difficult to grow or establish *e.g.* thebain from Papaver bracteatum,
- A source of novel chemicals in their own right *e.g.* rutacultin from culture of Ruta,
- As biotransformation systems either on their own or as part of a larger chemical process *e.g.* digoxin synthesis.

The Advantages of in Vitro Production of Secondary Metabolites

- The cell cultures and cell growth are easily controlled in order to facilitate improved product formation.
- The recovery of the product is easy.
- As the cell culture systems are independent of environmental factors, seasonal variations, pest and microbial diseases, geographical location constraints, it is easy to increase the production of the required metabolite.
- Mutant cell lines can be developed for the production of novel and commercially useful compounds.
- Compounds are produced under controlled conditions as per the market demands.
- The production time is less and cost effective due to minimal labour involved.

APPLICATIONS OF SECONDARY METABOLITES

Many of these secondary products especially various alkaloids are of immense use in medicine. The yield of these chemicals in cell culture, is though generally lower than in whole plants, it is substantially increased by manipulating physiological and biochemical conditions. Shikonine is a dye produced by the cells Lithospermum erythrorhizon on a commercial scale. Besides this there are a number of secondary metabolite products that are being widely used for various purposes. Vincristine is used as anticancer agent, digoxin controls cardiovascular disorders, pyrithrins is an insecticide etc. The production of specialty chemicals by plants has become a multibillion industry. Please refer to the table for some secondary metabolites and their uses.

Table. Showing Plant Species and Secondary Metabolites Obtained from them Using Tissue Culture Techniques

Product	Plant Source	Uses
Artemisin	Artemisia spp.	Antimalarial
Azadirachtin	Azadirachta indica	Insecticidal
Berberine	Coptis japonica	Antibacterial, anti

	inflammatory	
Capsaicin	Capsicum annum	Cures Rheumatic pain
Codeine	Papaver spp.	Analgesic
Camptothecin	Campatotheca accuminata	Anticancer
Cephalotaxine	Cephalotaxus harringtonia	Antitumour
Digoxin	Digitalis lanata	Cardiac tonic
Pyrethrin	Chrysanthemum	Insecticide (for grain
cinerariaefolium	storage)	
Morphine	Papaver somniferum	Analgesic, sedative
Quinine	Cinchona officinalis	Antimalarial
Taxol	Taxus spp.	Anticarcinogenic
Vincristine	Cathranthus roseus	Anticarcinogenic
Scopolamine	Datura stramonium	Antihypertensive

Production of Somatic Hybrids and Cybrids

The Somatic cell hybridization/parasexual hybridization or Protoplast fusion offers an alternative method for obtaining distant hybrids with desirable traits significantly between species or genera, which can not be made to cross by conventional method of sexual hybridization.

SOMATIC HYBRIDIZATION

Somatic hybridization broadly involves in vitro fusion of isolated protoplasts to form a hybrid cell and its subsequent development to form a hybrid plant.

The process involves:

- Fusion of protoplasts,
- Selection of hybrid cells,
- Identification of hybrid plants.

During the last two decades, a variety of treatments have been used to bring about the fusion of plant protoplasts. Protoplast fusion can be achieved by spontaneous, mechanical, or induced fusion methods..

These treatments include the use of fusogens like $NaNO_3$, high pH with high Ca^{2++} ion concentration, use of polyethylene glycol (PEG), and electrofusion.

These inducing agents used in protoplast fusion are called 'fusogen'. PEG treatment is the most widely used method for protoplast fusion as it has certain advantages over others.

These are:

- It results in a reproducible high-frequency of heterokaryon formation.,
- The PEG fusion is non specific and therefore can be used for a wide range of plants.,

- It has low toxicity to the cell and
- The formation of binucleate heterokaryons is low.

MECHANISM OF FUSION

The fusion of protoplasts takes place in three phases- agglutination, plasma membrane fusion and formation of heterokaryons. When the two protoplasts come in close contact with each other, they adhere to each other. This agglutination can be induced by PEG, high pH and high Ca^{2+}.

The protoplast membranes get fused at localized sites at the point of adhesion. This leads to the formation of cytoplasmic bridges between protoplasts. High pH and high Ca^{2+} ions neutralize the surface charges on the protoplasts which allows closer contact and membrane fusion between agglutinated protoplasts. The fused protoplasts become round as a result of cytoplasmic bridges which leads to the formation of spherical homokaryon or heterokaryon.

SELECTION OF HYBRID CELLS

The methods used for the selection of hybrid cells are biochemical, visual and cytometric methods using fluorescent dyes. The biochemical methods for selection of hybrid cells are based on the use of biochemical compounds in the medium. The drug sensitivity method is useful for the selection hybrids of two plants species, if one of them is sensitive to a drug. Another method, auxotrophic mutant selection method involves the auxotrophs which are mutants that cannot grow on a minimal medium. Therefore specific compounds are added in the medium.

The selection of auxotropic mutants is possible only if the hybrid cells can grow on a minimal medium. The visual method involves the identification of heterokaryons under the light microscope. In some of the somatic hybridizations, the chloroplast deficient protoplast of one plant species is fused with the green protoplast of another plant species. The heterokaryons obtained are bigger and green in colour while the parental protoplasts are either small or colourless. The cytometric method uses flow cytometry and flourescent-activated cell sorting techniques for the analysis of plant protoplasts.

APPLICATIONS OF SOMATIC HYBRIDIZATION

- *Creation of hybrids with disease resistance*: Many disease resistance genes (*e.g.* tobacco mosaic virus, potato virus X, club rot disease) could be successfully transferred from one species to another. *E.g* resistance has been introduced in tomato against diseases such as TMV, spotted wilt virus and insect pests.
- *Environmental tolerance*: Using somatic hybridization the genes conferring tolerance for cold, frost and salt were introduced in *e.g.* in tomato.

- *Cytoplasmic male sterility*: Using cybridization method, it was possible to transfer cytoplasmic male sterility.
- *Quality characters*: Somatic hybrids with selective characteristics have been developed *e.g.* the production of high nicotine content.

CHROMOSOME NUMBER IN SOMATIC HYBRIDS

The chromosome number in the somatic hybrids is generally more than the total number of both of the parental protoplasts. If the chromosome number in the hybrid is the sum of the chromosomes of the two parental protoplasts, the hybrid is said to be symmetric hybrid. Asymmetric hybrids have abnormal or wide variations in the chromosome number than the exact total of two species.

In 1972, Carlson and his associates produced the first inter-specific somatic hybrid between Nicotiana glauca and N. langsdorffii. In 1978, Melchers and his co-workers developed the first inter-genetic somatic hybrids between Solanum tuberosum (potato) and Lycopersicon esculentum (tomato). The hybrids are known as 'Pomatoes or Topatoes'.

LIMITATIONS OF SOMATIC HYBRIDIZATION

- Somatic hybridization does not always produce plants that give fertile and visible seeds.
- There is genetic instability associated with protoplast culture.
- There are limitations in the selection methods of hybrids, as many of them are not efficient.
- Somatic hybridization between two diploids results in the formation of an amphidiploid which is not favourable therefore haploid protoplasts are recommended in somatic hybridization.
- It is not certain that a specific character will get expressed in somatic hybridization.
- Regenerated plants obtained from somatic hybridization are often variable due to somaclonal variations, chromosomal elimination, organelle segregation etc.
- Protoplast fusion between different species/genus is easy, but the production of viable somatic hybrids is not always possible.

CYBRIDS

The cytoplasmic hybrids where the nucleus is derived from only one parent and the cytoplasm is derived from both the parents are referred to as cybrids. The process of formation of cybrids is called cybridization. During the process of cybridization and heterokaryon formation, the nuclei are stimulated to segregate so that one protoplast contributes to the cytoplasm while the other contributes nucleus alone. The irradiation with gamma rays and X-rays and use of metabolic inhibitors makes the protoplasts inactive and non-dividing. Some of the genetic traits in certain plants are cytoplasmically controlled. This includes certain types of male sterility, resistance to certain antibiotics and herbicides. Therefore cybrids are important for the transfer of cytoplasmic male sterility (CMS), antibiotic and herbicide resistance in agriculturally useful plants. Cybrids of Brassica raphanus that contain nucleus of B. napus, chloroplasts of atrazinc resistant B. capestris and male sterility from Raphanus sativas have been developed.

PLANT GERMPLASM CONSERVATION

Germplasm refers to the sum total of all the genes present in a crop and its related species.

The conservation of germplasm involves the preservation of the genetic diversity of a particular plant or genetic stock for it's use at any time in future. It is important to conserve the endangered plants or else some of the valuable genetic traits present in the existing and primitive plants will be lost. A global organization- International Board of Plant Genetic Resources (IBPGR) has been established for germplasm conservation and provides necessary support for collection, conservation and utilization of plant geneic resources through out the world.

The germplasm is preserved by the following two ways:

1. *In-situ conservation*: The germplasm is conserved in natural environment by establishing biosphere reserves such as national parks, sanctuaries. This is used in the preservation of land plants in a near natural habitat along with several wild types.
2. *Ex-situ conservation*: This method is used for the preservation of germplasm obtained from cultivated and wild plant materials. The genetic material in the form of seeds or in vitro cultures are preserved and stored as gene banks for long term use.

In vivo gene banks have been made to preserve the genetic resources by conventional methods *e.g.* seeds, vegetative propagules, etc. In vitro gene banks have been made to preserve the genetic resources by non - conventional methods such as cell and tissue culture methods. This will ensure the availability of valuable germplasm to breeder to develop new and improved varieties.

The methods involved in the in vitro conservation of germplasm are:

- *Cryopreservation*: In cryopreservation, the cells are preserved in the frozen state. The germplasm is stored at a very low temperature using solid carbon dioxide (at -790C), using low temperature deep freezers (at -800C), using vapour nitrogen (at- 1500C) and liquid nitrogen (at-1960C). The cells stay in completely inactive state and thus can be conserved for long periods. Any tissue from a plant can be used for cryopreservation *e.g.* meristems, embryos, endosperms, ovules, seeds, cultured plant cells, protoplasts, calluses. Certain compounds like- DMSO (dimethyl sulfoxide), glycerol, ethylene, propylene, sucrose, mannose, glucose, praline, acetamide etc are added during the cryopreservation. These are called cryoprotectants and prevent the damage caused to cells (by freezing or thawing) by reducing the freezing point and super cooling point of water.

- *Cold Storage*: Cold storage is a slow growth germplasm conservation method and conserves the germplasm at a low and non-freezing temperature (1-90C). The growth of the plant material is slowed down in cold storage in contrast to complete stoppage in cryopreservation and thus prevents cryogenic injuries. Long term cold storage is simple, cost effective and yields germplasm with good survival rate. Virus free strawberry plants could be preserved at 100C for about 6 years. Several grape plants have been stored for over 15 years by using a cold storage at temperature around 90C and transferring them in the fresh medium every year.
- *Low pressure and low oxygen storage*: In low- pressure storage, the atmospheric pressure surrounding the plant material is reduced and in the low oxygen storage, the oxygen concentration is reduced. The lowered partial pressure reduces the in vitro growth of plants. In the low-oxygen storage, the oxygen concentration is reduced and the partial pressure of oxygen below 50 mmHg reduces plant tissue growth. Due to the reduced availability of O_2, and reduced production of CO_2, the photosynthetic activity is reduced which inhibits the plant tissue growth and dimension. This method has also helped in increasing the shelf life of many fruits, vegetables and flowers.

4

Process of Plant Tissue Culture

Tissue culture is the process whereby small pieces of living tissue (explants) are isolated from an organism and grown aseptically for indefinite periods on a nutrient medium. For good success in plant tissue culture, it is good to start with explant that is meristems because these cells are capable of rapid proliferation. The used explants include buds, root tips, nodal segments or germinating seeds and these are placed on suitable culture media where they grow into an undifferentiated mass known as Callus. Because the nutrient media used in the technique support the growth of microorganisms, the explant is first washed with disinfectant such as sodium hydrochloride, hydrogen peroxide or meseuric chloride. Once established, the callus can be propagated indefinitely by sub-division. The nutrient media contains growth nutrient/substances and hormones, *i.e.*, auxin, cytokinin and gibberddin. The absolute amounts of hormones vary for different tissue, tissue explants from different parts of the same plant and for similar explants from different plants. There is no ideal medium, most of the in common use consist of inorganic salts, trace metals, vitamins, organic nitrogen sources, inositol, sucrose, and growth regulators. Organic nutrients cush as Casein hydrolysate or yeast extract and getting agent and optional extras. One of the composition of typical growth medium is given below.

Let us know something more about culturing of cells and plant organs:

- Plant Cell Culture: By transfer of callus culture to a liquid medium and agitated, the cell mass breaks to give a suspension of isolated cells with few small cell clusters and aggregates, the latter being extremely heterogenous. The cells can be subculture and maintained *in vitro* indefinitely. Genetic instability is added to the heterogenitity. Some plants like nicofiana tabacum (tobacco) and Glycine max (Soyabean) yield very fiable celluses and the cell lines obtained from these species are more homogenous.

When placed in a suitable medium is necessary. This medium is prepared by culturing high densities of the same or different species in fresh medium for few days and then removing the cells by filter sterilisation. The filterate contains many essential substances for growth. By placing singe cell on such medium, lines from single cells can be placed as in the case of micro organisms. Plant cells, however, instead of following colonies, proliferate in Callus. The occurrence of such single line of cells, different in characters, built from cell mutants is known as somatic variations.

Table. Composition of Murashige and Skoog (MS) Culture medium.

Ingredient	Amount (mg/l)
Sucrose	30000
$(NH_4)NO_3$	1650
KNO_3	1900
$CaCl_2.2H_2O$	440
$MgSO_4.7H_2O$	370
KH_2PO_4	170
$FeSO_4,4H_2O$	27.8
Na_2EDTA	37.3
$MNSO_4.4H_2O$	22.3
$ZnSO_4,7H_2O$	8.6
H_3SO_3	6.2
KI	0.83
$NaHO_4,2H_2O$	0.25
$CoCl_2.6H_2O$	0.025
Myo.inositol	100
Glyscine	2.0
Kinetin (a cytokinetic)	0.04 -10.0
Idolacetic acid	1.0 - 30.0

WHAT IS TISSUE CULTURE

The propagation of a plant by using a plant part or single cell or group cell in a test tube under very controlled and hygienic conditions is called "Tissue Culture".

Status in India

Banana is a globally important fruit crop with 97.5 million tones of production. In India it supports livelihood of million of people. With total annual production of 16.91 million tones from 490.70 thousand ha., with national average of 33.5 T/ha. Maharashtra ranks first in production with 60 T/ha. Banana contributes 37 per cent to total fruit production in India.

Banana is one of the major and economically important fruit crop of Maharashtra. Banana occupy 20 per cent area among the total area under crop in India. Maharashtra ranks second in area and first in productivity in India. Jalgaon is a major Banana growing district in Maharashtra which occupy

50,000 hectares area under Banana. But most of Banana is grown by planting suckers. The technology development in agriculture is very fast, it results in developing Tissue Culture Technique.

Agro Climate

Banana is basically a tropical crop, grows well in temperature range of 13ºC – 38ºC with RH regime of 75-85 per cent. In India this crop is being cultivated in climate ranging from humid tropical to dry mild subtropics through selection of appropriate varieties like Grandnaine. Chilling injury occurs at temperatures below 12ºC. The normal growth of the banana begins at 18ºC, reaches optimum at 27ºC, then declines and comes to a halt at 38ºC. Higher temperature causes sun scorching. High velocity wind which exceeds 80 km phrs damages the crop.

Soil

Soil for banana should have good drainage, adequate fertility and moisture. Deep, rich loamy soil with pH between 6-7.5 are most preferred for banana cultivation. Ill drained, poorly aerated and nutritionally deficient soils are not suitable for banana. Saline solid, calcareous soil are not suitable for Banana cultivation. Avoided soil of low laying areas, very sandy and heavy black cotton with ill drainage.

A soil that is not too acidic and not too alkaline, rich in organic material with high nitrogen content, adequate phosphorus level and plenty of potash are good for banana.

Varieties

In India banana is grown under diverse conditions and production systems. Selection of varieties, therefore is based on a large number of varieties catering to various kinds of needs and situations. However, around 20 cultivars, *viz.* Dwarf Cavendish, Robusta, Monthan, Poovan, Nendran, Red banana, Nyali, Safed Velchi, Basarai, Ardhapuri, Rasthali, Karpurvalli, Karthali and Grandnaine, etc.

Grandnaine is gaining popularity and may soon be the most preferred variety due to its tolerance to biotic stresses and good quality bunches. Bunches have well spaced hands with straight orientation of figures, bigger in size. Fruit develops attractive uniform yellow colour with better self life and quality than other cultivars.

Land Preparation

Prior to planting banana, grow the green manuring crop like daincha, cowpea, etc., and burry it in the soil. The land can be ploughed 2-4 times and levelled. Use ratovator or harrow to break the clod and bring the soil to a fine tilt. During soil preparation basal dose of FYM is added and thoroughly mixed into the soil. The furrows to be open with help tractor for planting or prepared raised bed in heavy soil. Apply 20 kg of FYM (well decomposed), 250 gm of Neem cake and 20 gm of conbofuron per plant. Prepared pits are left to solar radiation helps in killing the harmful insects, is effective against soil borne diseases and aids aeration. In saline alkali soil where PH is above 8 Pit mixture is to be modified to incorporate organic matter.

Addition of organic matter helps in reducing salinity while addition of purlite improves, porosity and aeration. Alternative to planting in pits is planting in furrows. Depending on soil strata one can choose appropriate method as well as spacing and depth at which plant is required to be planted.

THEORY OF TISSUE CULTURE

The theoretical basis for plant tissue culture was proposed by Gottlieb Haberlandt, German Academy of science in 1902 on his experiments on the culture of single cell. The first true cultures were obtained by Gautheret from cambial tissue of Acer pseudoplatanus. The term plant tissue culture (Micro propagation) is generally used for the aseptic culture of cells, tissues, organs and their components under defined chemical and physical conditions *in vitro*. The basic concept of the plant body can be dissected into smaller part termed as "explants" and any explants can be developed into a whole plant. It is a central innovative areas of applied plant science, including agriculture and plant biotechnology.

This technique is effective because almost all the plants cell are totipotent; In each cell possesses the genetic information and cellular machinery necessary to generate the whole organism. Since, this technique can be used to produce a higher number of plants that are genetically similar to a parent plant as well as to another. Two concepts, plasticity and totipotency, are the central processes to understand the regeneration and plant cell culture. Plants, due to its longer life span and sessile nature, have developed a greater ability to overcome the extreme conditions.

Most of the processes include in plant development and the growth, adapt to environmental conditions. When the plant cells and tissues are cultured *in vitro*, most of them are generally exhibit a very high degree of plasticity, which allows one type of organ or tissue to be initiated from another type. Like this way, the whole plant can be subsequently regenerated. These maintenance of genetic potential is called totipotency.

The plant tissue culture medium is an artificial nutrient supplement of organic and inorganic nutrients used for cultivation of plant tissue media. The appropriate composition of the medium largely determines the success of the culture. The culture media used for the *in vitro* cultivation of the plant cells are composed of three basic components.

- Essential elements (normal ions) supplied as a complex mixture of salts.

- An organic supplements providing vitamins and amino acids.
- A source of fixed carbon which is usually supplied as sucrose.

When cultured in an appropriate medium having auxin and cytokinin, explants will give rise to an unorganised, growing and dividing mass of cells called callus. Callus cultures are initiated from a small part of an organ or tissue segment called the explants on a growth supporting solidified nutrient medium under sterile conditions.

Any part of the plant organ or tissues may be used as the explants. At the time of callus formation, there is some degree of dedifferentiation happens both in morphology and metabolism. One of the major consequences of this dedifferentiation is that most plant cultures lose their ability to perform photosynthesis. The necessitates of the addition of other components such as carbon and vitamins source to the culture media, in addition to the unusual mineral nutrients.

MORPHOLOGY OF CALLUS

Callus varies considerably in appearance and texture, ranging from hard nodular cell masses to friable soft ones. They maybe white or creamish, orange, green either in whole or part as a result of chloroplast development. The shape of individual cells within the callus mass ranges from the near spherical or markedly elongated. A typical unorganised plant callus initiated from a new explants or piece of previously initiated calli has three stages of development.

1. The induction of cell division.
2. A period of active cell division during which differentiated cells lose specialised features they may have acquired and become dedifferentiated. Cell division usually occurs in the outer layer of the explants.
3. Period when cell division slows down on ceases and when within the callus, there is increasing cellular differentiate.

Callus culturing is performed in the dark while light can be encourage the differentiation of the callus. At the time of long term culture, the culture may loss the requirement for cytokinin and auxins. Manipulation of the auxins to cytokinin ratio in the medium can leads to the development of shoots, roots or somatic embryos from which the plant can be subsequently produced.

Callus culture is useful for many purposes:

- Callus is the starting material for the suspension culture which cells are separated.
- It helps in the production of secondary plant products.
- It is useful for the synthesis of starting compounds that are subsequently modified to yield the desired product.
- It is the starting materials for vegetative propagation of plants.

Based on the availability of the various *invitro* techniques, the dramatic increase in their application to various problems in basic biology, agriculture, horticulture, and forestry.

The applications can divide conveniently into six broad areas:

1. Cell behaviour.
2. Plant modification.
3. Germplasm storage and pathogen –free plants.
4. Clonal propagation.
5. Product formation.
6. Improved varieties.

PROTOPLASTS CELLS

These are cells minus their cell walls which form useful material for cell manipulations as under certain conditions, contrasting cell types can be fused to yield somatic hybrids; a process known as protoplast fusion. Protoplasts can be produced from suspension cultures, callus or intact tissues by mechanical disruption or treatment with enzymes.

Pectinase breaks cell aggregates into individual cells and cellulose removes the cell wall from these protoplast suspension, by washing and plating on nutrient medium, protoplasts form new cell wall in 5-10 days and initiate cell division. Plant Organ culture: This is the most common type known as Tissue Culture. Here shoot tips are after surface sterilisation are placed in growth medium lacking hormones where these develop into single seedling like shoots.

If instead cytokinin is used in the medium, axillary shoots will emerge and produce a shoot culture. These can be further subdivided into smaller clumps or separate shoots, which will in turn form similar clusters when subcultured on fresh medium. Provided basic nutrient formulation is adequate for normal growth this subdivision process known as micropropogation can be undertaken every 4-8 weeks for a quite long time say up to 10-12 sub-cultures in case of banana.

Micropropogation indefinite cultures of plant roots can be achieved in the similar manner, with use of auxin in the medium and no cytokinnin many commercial companies are using micropropogation of crop plants like Banana, Sugarcane, spices crops and Ornamentals like carnation, gerbera, and antherium. Uniformity of quality, free from virus infection, high yield, etc., are some of the advantages of micropropagation of commercial crops. Since the process is laborious and time taking in many countries, the same has been fully automated.

Regeneration of Plants: Numerous species of plants can be maintained in cell or callus culture. In this state new genotypes can be isolated by somaclonal variation; protoplast fusion, or mutagenesis. These cells to be useful must be converted into an intact plant. The cells capable of developing into intact plants are said to be totipotent; totipotency being the property of undetermined cells. Undetermined cells can be defined as those, with capacity to switch to different development pathways depending on environment in which they grow undetermined cells most readily develop into a callus. It should be remembered that for callus formation auxin and cytokinin both are required; whereas only a cytokinin both are required; whereas only a cytokinin is required for shoot culture and only an auxin for root culture.

Increasing levels of Cytokinin or auxin in a culture, shoot or root formation can be induced and promoted in culture. The formation of roots or shoots in culture is known as Organogenesis. Under certain cultural conditions callus can be induced to undergo different development process known as somatic embryogenesis.

This is a pattern of differentiation, similar to that seen in zygotes after fertilization, to produce embryos. These cells are embryo like but produced from somatin cells, and not from fusion of two germ cells. These embryos can develop into fully functional plants without the need to induce root and shoot formation on artificial media Embryos formation is stimulated by transferring callus grown in media containing auxin. 2,4-d in to a 2,4-D free media containing reduced sources of on nitrogen, *i.e.*, ammonium salts.

Anther and Pollen culture: Pollen grains form pollen tube and male gametes, when placed on suitable nutrient media like MS, most pollen grain follow normal development, but a few grains will form a callus instead. In a similar manner, anthers can be cultured to form embryos directly from pollen grains.

These embryos can be induced to develop true haploid plants. With optimal nutrient conditions for donor plants and explants are provided, it is possible to obtain several hundred haploid plants from a single anther. For successful anther culture if is necessary to excise flower buds at first mitotic division in the uninucleated microspores tetrads. Other treatments like growing on different nutrient media, cold storage, for few days to several weeks, etc., can increase the frequency of successful embryos.

Application of cell, tissue and organ culture:

- Production of fine chemicals like Codeine, Diosgenin, Sitostero, Quinine, Vincristine, Atropine, Pyrethrin, Saffoon and Methanol which are presently produced from plant sources can be produced from cell culture. However, the commercial success has been obtained in production of 'Shikonin' from cells of plant lithospermusm erythorhizon.
- *Production of pathogen -free plants:* Crops and ornamental plants can be infected by a wide variety microbial pests; Viroids, viruses, mycoplasms, bacteria, fungi and nematodes. The infections greatly affect yield and quality of plants. In some cases, infection can be eliminated by micropropagation of unaffected parts of the plants. The basic method of obtaining virus-free plants is culture of apical meristems. If a small enough piece is cultured preferably first half

mm or so, of the growing tip, there is possibility to obtain virus free plants which can be further multiplied. There are other methods like high temp treatment, incubation of explants on suitable culture media containing malachite green or thiouracil which can reduce virus infection and replication.

- *Large scale plant propagation:* In certain situation large scale vegetative multiplication of crop plants/fruit trees/ ornamentals is useful to save time get uniform quality and disease free plants. Such plants can be obtained by
  - *In vitro* methods of propogation through callus and cell culture as in oil palm
  - Production of minituber
  - Micropropagation as in case of Banana.
  - Germplasm conservation through seed storage and multiplication has many problems. The current method of germplasm maintenance is being replaced by methods like cell, tissue and organ culture. In this methods, the advantages are
    a. Extremely slow growth involving depression of metabolism and]
    b. Slow growth which is obtained by cryopreservation of cell and organ culture. However, the basic requirement of gene bank or germplasm preservation is that original characters of the genotype should not be changed. This can not be guaranteed in *invitro* preservation, plant cell, organ and tissue culture, are all developing technologies. All the three has a role to play in academic as well as in applied often commercial fields. Plant tissue culture has already proved to be a regular technique for propagation and quick multiplication of many crop plants as well as ornamentals. The success and performance of these techniques are dependent on many factors like development of fool-proof techniques to suit the needs of a plant variety, cost of production -skills of operators, and continuos research support for improving the techniques and finally rigorous testing of the micropropagated plants for their being free from diseases and also for their yield and quality. All these need a time-bound research and development programme having the involvement of private and public sector.

All these culture techniques not only are related to plant bio-technology but often form the part of it. More we get involved in exploiting these techniques on commercial scale, more we will be close to harvest the advantages of various aspects of plant biotechnology.

FUNCTION OF TISSUE CULTURE

In the recent years plant protoplast, cell and tissue cultures have become an important tool for crop improvement, commercial production of natural compounds and in the development of forestry.

Some of the areas of biotechnological application of cultured plant protoplasts/ cells/tissues are:

- Tissue culture applications in order to capitalise on the totipotency of cells,
- Cells and protoplast cultures coupled 'with DNA vectors to overcome problems caused by barriers to gene transfer through sexual means,
- Culture of plant cells for the production of useful compounds,
- Extension and increase of deficiency of biological nitrogen fixation, and
- Transfer of genes for nitrogen fixation ability to non-fixing species.

Due to the manipulation of plant tissues in the laboratory this technique has been referred to by some researchers, as a 'botanical laser' whose numerous uses are yet to be fully understood.

APPLICATION IN AGRICULTURE

The major areas of application of tissue culture are briefly highlighted here.

Improvement of Hybrids

Development of cell fusion and hybridisation techniques have solved the problem of incompatibility of plants and widened the scope of production of new varieties with in a short time. For example breeders obtained somatic hybrid of wild and cultivated potatoes (*S. tuberosum, S. chacoense*) and succeeded in the induction or organogenesis.

The somatic hybrid plant inherited many characters, *viz.*, intermediate leaf morphology, stomata, forms and colour of tubers prolonged flowering, large and fertile pollen grains, high yield, resistance against Y-virus. However, haploid plant materials available as protoplast, cell and tissue culture systems are currently being evaluated for the use in transfer of foreign genetic material to selected plant species by protoplast fusion, transformation, transduction and organelle transfer.

Production of Encapsulated Seeds

The two 'techniques, somatic embryogenesis and organogenesis are the alternative methods for regeneration of the whole plant from cultured tissue *in vitro*. Moreover, the recent works on the wrapping an *in vitro* derived embryos in a seed coat, *i.e.*, production of synthetic seeds are important because it may improve agriculture due to its low cost.

Production of Disease Resistant Plants

Many plant species, which propagate vegetatively are systematically infected by virus, bacteria, fungi and nematodes. Their inoculum is carried over several generations resulting in continued adverse effect of productivity and quality of crops. In order to ensure highest possible yield and quality, it is necessary to provide disease free stock plants to growers. Tissue culture techniques have solved this problem and minimized the time of biological testing.

Unless large scale population of pure inoculum of test pathogens are available, it is difficult to pursue the establishment of pathogenicity and crop loss-assessment as it is done in field conditions. Now it has become possible to carry out such experiments in laboratory within short span of time by using tissue culture technologies. Scientists have discussed the following advantages for the study of several aspects of host-pathogen interactions and responses.

They are:

- Ability to isolate host cells without wounding,
- Control of inoculum of pathogen and number of host cells,
- Ability t'J change the nature of host pathogen interaction by altering the constituent of growth medium,
- Presence of only one or a few major host cell types and
- Easy to apply and remove materials, *e.g.*, labelled precursors form cultured cells.

Production of Stress Resistance Plant

Biochemical mechanisms exist is Culture cells which determine the resistance to biocide chemicals and provide the theoretical promise for selection *in vitro*. For example: cell suspension tolerant to 2,4-D when the cells were subcultured for six months in liquid medium supplemented with increasing amount of herbicides. The cells were able to grow in one rnM (milli mol) 2,4-D, while the control suspension was completely inhibited at O.3mM 2,4-D. It is suggested that tolerance was the result of induction of enzymatic systems responsible for the degradation of 2,4-D.

Uses of protoplasts as a system to select cell lines tolerant to herbicides have not been extensively explored. Protoplast technology can be used to increase the possibility to obtain monoclonal lines and offer the opportunity for intraspecific\transfer of cytoplasmic factor of resistance to some type of herbicide. A potential application of transfer of herbicides tolerance factor is the transplant of cytoplasmic organelles, for l example chloroplasts.

Transfer of Nif Gene to Eukaryotes

Nitrogen fixing ability, a genetic character, exists in prokaryotic diazotrophs. However, one of the major tasks is the transfer of this character to eukaryotes. In recent years, researches are being done to solve this problem through tissue culture techniques coupled with the recombinants DNA technology. Historically, nitrogen fixation by rhizobia was believed to occur through symbiosis. For the first time an excitement was caused in scientific community with the discovery by Holsten *et al.* (1971). They obtained active rhizobia in the absence of nodules, leghaemoglobin and bacteroids which were apparently necessary in the intact plants.

They established *Rhizobium joponicum.* In cell suspension of soybean roots. Callus induced from root explants of soybean on a specific medium was inoculated with R. joponicum. Later on it was microscopically observed that infection threads were formed by the bacteria which were present between intracellular spaces. They multiplied inside cells. Moreover, development of nitrogenase in soybean callus- Rhizobium system growing on solid was also observed and it was also found that only specific (isomorphic) cells of callus are vulnerable to infection by the bacteria. In addition to improvement the bacterial strains and increased nodulation, it is necessary to seek those genotypes with the efficient photosynthesis and improved partitioning of carbohydrates to nodules.

Future Prospects

In addition to work done successfully on nif gene transfer, there are other important genes which have been clones, for example:

- Phaseolin and leghaemoglobin genes in soyabean;
- Storage protein genes in soybean,
- Genes of ribulose bi-phosphate carboxylase/oxygenase (Ru BP case) of pea, maize, wheat, etc.

Success achieved on these aspects would certainly promote in green revolution. Moreover, improvement in primary productivity by conversion of C_3 plants into C_4 ones through genetic engineering techniques hopefully would increase the primary productivity.

STRUCTURE AND FUNCTION OF PLANT TISSUE SYSTEMS

Although the architecture varies considerably among roots, stems, and leaves, all plant organs are comprised of 3 basic tissue types: dermal tissue, vascular tissue, and ground tissue. A tissue is simply an assemblage of cells that work together as a structural or functional unit. In this lab, we will examine the general arrangements of those tissues among different plant organs.

You will find that the distribution of tissues varies somewhat among different plants as well as among different organ systems. For example, one of the key differences between monocots and dicots has to do with the arrangement of vascular tissues in leaves and stems. In next week's lab, we will continue our examination of plant tissues by exploring some of the different cell types that form these three plant tissue sytems.

Dermal Tissue

The dermal tissue system is derived from the protoderm and consists of 2 tissue types*:

- *Periderm*: Periderm includes cork and cork-producing tissues and is only found in plants with secondary growth.
- *Epidermis*: The epidermis constitutes the outmost layer of cells of roots, stems and leaves. Epidermal cells are usually covered by a waxy layer, called a cuticle that serves for waterproofing the cells.

Vascular Tissue

Vascular tissues, derived from the procambium*, are the tissues responsible for transporting substances from one point in the plant body to another. As is the case with animals, the vascular tissues are responsible for the movement of dissolved gases, sugars, and minerals throughout the organism.

In plants, there are 2 types of conducting tissues:

- *Xylem*: Xylem is the tissue specialised for the conduction water and dissolved minerals. Cells of the xylem are dead at maturity and act as the structural medium for the movement of water. Individual xylem cells have multiple holes in the walls to allow for the passage of water between adjacent cells. The bulk of wood consists of xylem tissue.
- *Phloem*: Phloem conducts dissolved sugars throughout the plant. The main conducting cells of phloem, called"sieve elements", are alive at maturity, but are highly modified for carrying out the function of transport. For example, many of the organelles you might expect to find in a living plant cell, such as nuclei, ribosomes, Golgi apparati, cytoskeleton, and vacuoles are absent from the sieve elements in order to make more space for fluid transport. The small amount of protoplast in sieve elements is distributed in a thin layer along the periphery of the cell wall. In some cases, you may be able to observe smaller cells called"companion cells" distributed among the sieve elements. Companion cells contain all the organelles you would expect to find in a typical plant cell and carry out many of the necessary life functions for the sieve elements which lack organelles.

Ground Tissue

Derived from the ground meristem*, ground tissues are distributed between and among the dermal and vascular tissues of roots, stems, and leaves. Ground tissues include a wide variety of tissue types that perform a multitude of functions within the plant.

Some of the more familiar ground tissues include the following:

- *Cortex*: the ground tissue distributed between the epidermis and vascular tissues in roots and stems. Cortex is usually modified for storage, especially of starches. In herbaceous green stems, cortex often contains chloroplasts and thus is further modified for carrying out photosynthesis.
- *Pith*: like cortex, pith is generally modified for starch storage, but is only found to the interior of the vascular tissues in stems and roots.
- *Mesophyll*: the ground tissue of leaves found between the two layers of epidermis. Leaf mesophyll is the tissue most adapted for photosynthesis and thus is generally rich with chloroplasts.

In addition to the functions indicated above, ground tissues also have an important role in wound healing and the regeneration of damaged or dead cells throughout the plant.

PLANT TISSUES ARE DERIVED FROM MERISTEMS

Unlike animals, most plants grow (increase the number and size of their cells) throughout their lifetime. This ability to grow indeterminately is due to a special form of tissue known as a"meristem". A meristem is a region of tissue that is perpetually embryonic-that is, it is relatively undifferentiated and divides continuously, giving rise to all new cells and tissues that make up the plant body.

Can you think of an analogy to the plant's meristem in animals/humans? There are several meristems distributed throughout the plant body which are named for their location and/or for the tissue systems that are derived from them.

Apical (Primary) Meristems

Are located in the tips (or, apex) of the growing roots and shoots of all plants. Primary meristems produce primary tissues and account for all the primary growth of plants. Primary growth is growth that contributes to elongation of the plant (height and root lengthening) and for the establishment of its basic organ structure (roots, stems, leaves).

Lateral (Secondary) Meristems

Are organised as concentric cylinders (think of a can inside a can inside a can) within the plant body and are responsible for secondary growth. Secondary growth contributes to the thickening of a plant structure, or increases in girth. With few exceptions, lateral meristems tend to be present only in woody plants, such as trees, shrubs, and some vines. As cells in the apical meristem divide and give rise to new cells, the tips of roots and shoots extend outward, away from the central plant body. As a consequence of this elongation due to the addition of new cells, the slightly older meristematic cells become distributed into different regions within the root and shoot. These slightly more mature cells still retain their meristematic properties, but are somewhat differentiated from the true apical meristem. These regions of meristematic tissue derived from the apical meristem are referred to as"primary meristems" and are named for the tissues that they will give rise to:

- *Protoderm*: Gives rise to dermal tissues,
- *Ground Meristem*: Gives rise to ground tissues, and
- *Procambium*: Gives rise to vascular tissue (aka,"cambial tissue").

The organisation of primary meristems also varies among plant types and organs.

Like other organisms, the cells in a plant are grouped together into various tissues. These tissues can be simple, consisting of a single cell type, or complex, consisting of more than one cell type.

DERMAL TISSUE SYSTEM

The dermal tissue system consists of the epidermis and the periderm. The epidermis is a single layer of closely packed cells.

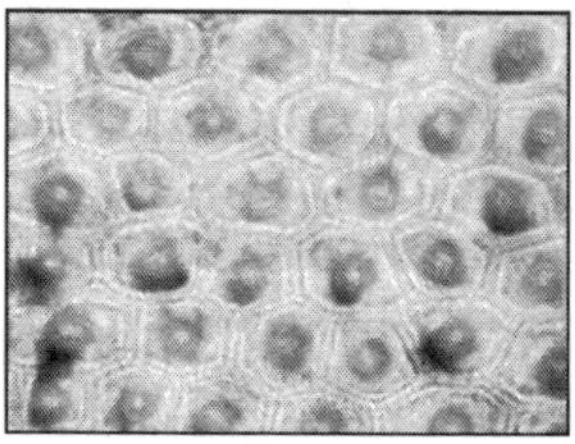

It both covers and protects the plant. It can be thought of as the plant's"skin." Depending on the part of the plant that it covers, the dermal tissue system can be specialised to a certain extent. For instance, the epidermis of a plant's leaves secretes a coating called the cuticle that helps the plant retain water.

Epidermis Cells

The periderm, also called bark, replaces the epidermis in plants that undergo secondary growth. The periderm consists of cork cells and protects the plant from pathogens, prevents excessive water loss and provides insulation for the plant.

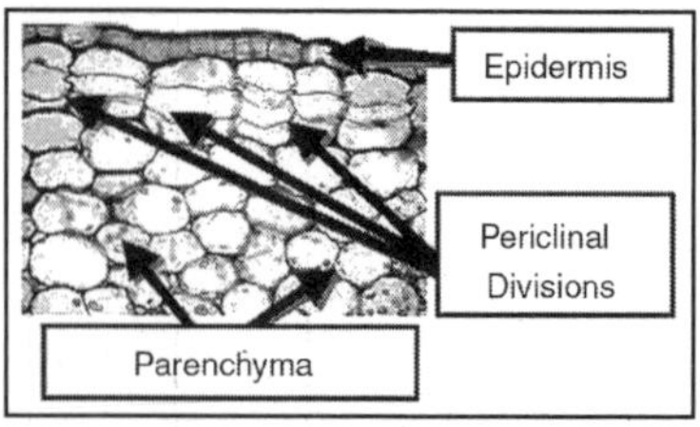

Fig. Periderm Formation is Indicated by Periclinal Divisions in Parenchyma cells.

Vascular Tissue System

Xylem and phloem throughout the plant make up the vascular tissue system. It allows water and other nutrients to be transported throughout the plant.

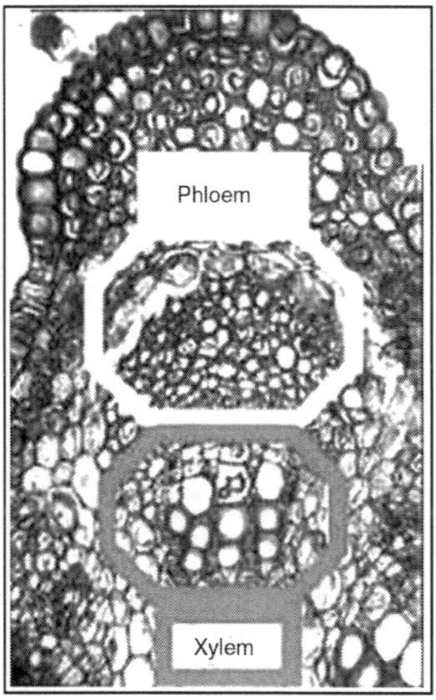

Fig. Xylem and Phloem Image

GROUND TISSUE SYSTEM

The ground tissue system synthesises organic compounds, supports the plant and provides storage for the plant. It is mostly made up of parenchyma cells but can also include some collenchyma and sclerenchyma cells as well.

PLANT TISSUE SYSTEMS: DERMAL, GROUND, AND VASCULAR

Dermal tissue. The epidermis is the dermal tissue of young plants undergoing primary growth. It is generally composed of specialized, flattened polygonal cells that occur on all plant surfaces. Shoot surfaces are usually coated with a waxy cuticle to prevent water loss and are often covered with hairs, or trichomes, which are epidermal cell extensions. Pairs of specialized epidermal cells, the guard cells, are found surrounding microscopic pores in all leaves. The guard cells and pores are called stomata (singular stoma), and they permit gas exchange (water loss, CO_2 uptake, and O_2 release or uptake) between the atmosphere and the interior of the leaf.

The root epidermis is adapted for absorption of water and minerals, and its outer wall surface typically does not have a waxy cuticle. Extensions from the root epidermal cells, the root hairs, increase the surface area over which absorption can take place. Ground tissue. Making up the bulk of the plant are cells termed the ground tissue. There are three types of ground tissue: parenchyma, collenchyma, and sclerenchyma.

- Parenchyma, the most abundant ground tissue, consists of thin-walled, metabolically active cells that carry out a variety of functions in the plant including photosynthesis and storage
- Collenchyma tissue is composed of narrow, elongated cells with thick primary walls. Collenchyma cells provide structural support to the growing plant body, particularly shoots, and their thickened walls are nonlignified so they can stretch as the organ elongates. Collenchyma cells are typically arranged in bundles or layers near the periphery of stems or leaf petioles.
- Sclerenchyma consists of two types of cells, sclereids and fibers. Both have thick secondary walls and are frequently dead at maturity. Sclereids occur in a variety of shapes, ranging from roughly spherical to branched, and are widely distributed throughout the plant. In contrast, fibers are narrow, elongated cells that are commonly associated with vascular tissues. The main function of sclerenchyma is to provide mechanical support particularly to parts of the plant that are no longer elongating.

In the stem, the pith and the cortex make up the ground tissue. The pith is located within the cylinder of vascular tissue, where it often exhibits a spongy texture because of the presence of large intercellular air spaces. If the growth of the pith fails to keep up with that of the surrounding tissues, the pith may degenerate, producing a hollow stem. In general, roots lack piths, although there are exceptions to this rule. In contrast, the cortex, which is located between the epidermis and the vascular cylinder, is present in both stems and roots.

At the boundary between the ground tissue and the vascular tissue in roots, and occasionally in stems, is a specialized layer of co rtex known as the endodermis. This single layer of cells originates from cortical tissue at the innermost layer of the root cortex and forms a cylinder that surrounds the central vascular tissue, or stele. Early in root development, a narrow band composed of the waxy substance suberin is formed in the cell walls circumscribing each endodermal cell.

These suberin deposits, called Casparian strips, form a barrier in the endodermal walls to the intercellular movement of water, ions, and other water-soluble solutes to the vascular cells. Leaves have two interior layers of ground tissue that are collectively known as the mesophyll. The palisade parenchyma consists of closely spaced, columnar cells located beneath the upper epidermis. There is usually one layer of palisade parenchyma in the leaf. Palisade parenchyma cells are rich in chloroplasts and are a primary site of photosynthesis in the leaf. Below the palisade parenchyma are i rregularly shaped, widely spaced spongy mesophyll cells. The spongy mesophyll cells are also photosynthetic, and the large spaces between these cells allow diffusion of carbon dioxide. The spongy mesophyll also contributes to leaf flexibility in the wind, and this flexibility facilitates the movement of gases within the leaf.

Vascular tissues: xylem and phloem. The vascular tissue is composed of two major conducting systems: the xylem and the phloem. The xylem transports water and mineral ions from the root to the rest of the plant. The phloem distributes the products of photosynthesis and a variety of other solutes throughout the plant. The tracheids and vessel elements are the conducting cells of the xylem.

Both of these cell types have elaborate secondary-wall thickenings and lose their cytoplasm at maturity; that is, they are dead when functional. Tracheids overlap each other, whereas vessel elements have open end walls and are arranged end to end to form a larger unit called a vessel. Other cell types present in the xylem include parenchyma cells, which are important for the storage of energy-rich molecules and phenolic compounds, and sclerenchyma fibers.

The sieve elements and sieve cells are responsible for sugar translocation in the phloem. The former are found in angiosperms; the latter perform the same function in gymnosperms. Like vessel elements, sieve elements are often stacked in vertical rows, forming larger units called sieve tubes, whereas sieve cells form overlapping arrays. Both types of conducting cells are living when functional, but they lack nuclei and central vacuoles and have relatively few cytoplasmic organelles.

Substances are translocated from sieve cell to sieve cell laterally through circular or oval zones containing enlarged pores, called sieve areas. In contrast, sieve tubes translocate substances through large pores in the end walls of the sieve elements, called sieve plates. Sugar movement through sieve tubes is more efficient and rapid than through sieve cells and represents a more evolutionarily advanced mechanism. Sieve elements are associated with, and depend on, densely cytoplasmic parenchyma cells called companion cells. The analogous cells adjacent to the sieve cells of gymnosperms are called albuminous cells. Companion cells provide proteins and metabolites necessary for the functions of the sieve tube elements. In addition, the phloem frequently contains storage parenchyma and fibers that provide mechanical support.

THE STRUCTURES OF CHLOROPLAST GLYCOSYLGLYCERIDES

The structures of the three plant glycosylglycerides, monogalactosyldiacylglycerol (MGDG), digalactosyldiacylglycerol (DGDG) and sulfoquinovosyldiacylglycerol (SQDG), are shown in Figure. The two galactolipids, MGDG and DGDG, are uncharged but polar; the sulfolipid, SQDG, is negatively charged as well as being polar (Harwood 1997).

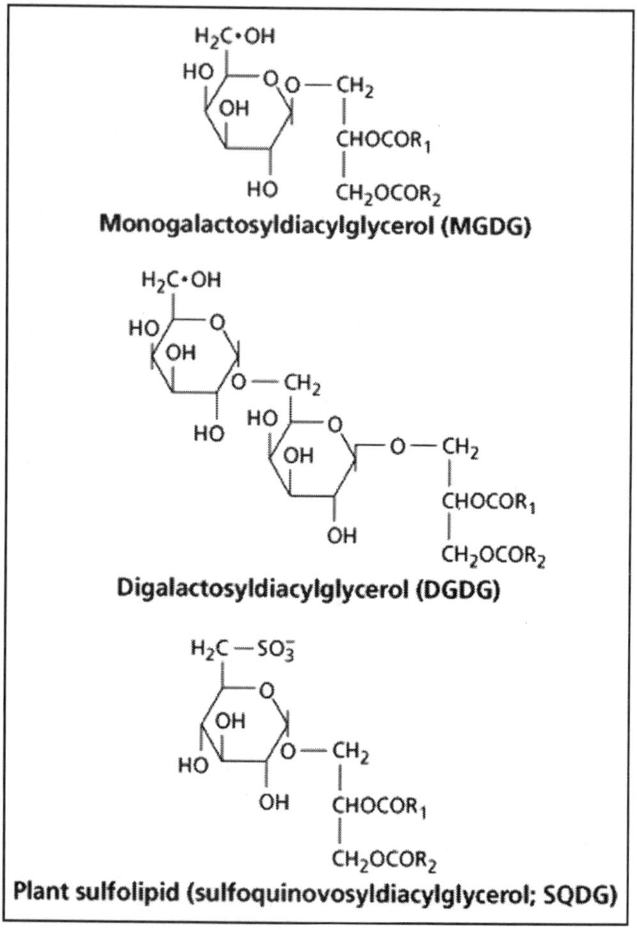

Fig. The Structures of Chloroplast Glycosylglycerides.

MGDG and DGDG are galactolipids; SQDG is a sulfolipid. Note that the two galactolipids are uncharged, while the sulfolipid is charged.

PLANT GROWTH REGULATORS AND TISSUE CULTURE

Generalisations about plant growth regulators and their use in plant cell culture media have been developed from initial observations made in the 1950s. There is, however, some considerable difficulty in predicting the effects of plant growth regulators: this is because of the great differences in culture response between species, cultivars and even plants of the same cultivar grown under different conditions.

However, some principles do hold true and have become the paradigm on which most plant tissue culture regimes are based. Auxins and cytokinins are the most widely used plant growth regulators in plant tissue culture and are usually used together, the ratio of the auxin to the cytokinin determining the type of culture established or regenerated. A high auxin to cytokinin ratio generally favours root formation, whereas a high cytokinin to auxin ratio favours shoot formation. An intermediate ratio favours callus production.

Intermediate ratio (callus formation) | Low auxin to cytokinin ratio (shoot formation) | High auxin to cytokinin ratio (root formation)

Fig. The effect of different ratios of auxin to cytokinin on the growth and morphogenesis of callus. High auxin to cytokinin ratios promote root development, low ratios promote shoot development. Intermediate ratios promote continued growth of the callus without differentiation.

CULTURE TYPES

Cultures are generally initiated from sterile pieces of a whole plant. These pieces are termed 'explants', and may consist of pieces of organs, such as leaves or roots, or may be specific cell types, such as pollen or endosperm. Many features of the explant are known to affect the efficiency of culture initiation. Generally, younger, more rapidly growing tissue (or tissue at an early stage of development) is most effective. Several different culture types most commonly used in plant transformation studies will now be examined in more detail.

Callus

Explants, when cultured on the appropriate medium, usually with both an auxin and a cytokinin, can give rise to an unorganised, growing and dividing mass of cells. It is thought that any plant tissue can be used as an explant, if the correct conditions are found. In culture, this proliferation can be maintained more or less indefinitely, provided that the callus is subcultured on to fresh medium periodically.

During callus formation there is some degree of dedifferentiation (*i.e.*, the changes that occur during development and specialisation are, to some extent, reversed), both in morphology (callus is usually composed of unspecialised parenchyma cells) and metabolism. One major consequence of this dedifferentiation is that most plant cultures lose the ability to photosynthesise.

This has important consequences for the culture of callus tissue, as the metabolic profile will probably not match that of the donor plant. This necessitates the addition of other components—such as vitamins and, most importantly, a carbon source—to the culture medium, in addition to the usual mineral nutrients. Callus culture is often performed in the dark (the lack of photosynthetic capability being no drawback) as light can encourage differentiation of the callus. During long-term culture, the culture may lose the requirement for auxin and/or cytokinin.

This process, known as 'habituation', is common in callus cultures from some plant species (such as sugar beet). Callus cultures are extremely important in plant biotechnology. Manipulation of the auxin to cytokinin ratio in the medium can lead to the development of shoots, roots or somatic embryos from which whole plants can subsequently be produced. Callus cultures can also be used to initiate cell suspensions, which are used in a variety of ways in plant transformation studies.

Cell-Suspension Cultures

Callus cultures, broadly speaking, fall into one of two categories: compact or friable. In compact callus the cells are densely aggregated, whereas in friable callus the cells are only loosely associated with each other and the callus becomes soft and breaks apart easily. Friable callus provides the inoculum to form cell-suspension cultures. Explants from some plant species or particular cell types tend not to form friable callus, making cell-suspension initiation a difficult task. The friability of callus can sometimes be improved by manipulating the medium components or by repeated subculturing. The friability of the callus can also sometimes be improved by culturing it on 'semi-solid' medium (medium with a low concentration of gelling agent).

When friable callus is placed into a liquid medium (usually the same composition as the solid medium used for the callus culture) and then agitated, single cells and/or small clumps of cells are released into the medium. Under the correct conditions, these released cells continue to grow and divide, eventually producing a cell-suspension culture. A relatively large inoculum should be used when initiating cell suspensions so that the released cell numbers build up quickly. The inoculum should not be too large though, as toxic products released from damaged or stressed cells can build up to lethal levels.

Large cell clumps can be removed during subculture of the cell suspension. Cell suspensions can be maintained relatively simply as batch cultures in conical flasks. They are continually cultured by repeated subculturing into fresh medium. This results in dilution of the suspension and the initiation of another batch growth cycle. The degree of dilution during subculture should be determined empirically for each culture. Too great a degree of dilution will result in a greatly extended lag period or, in extreme cases, death of the transferred cells. After subculture, the cells divide and the biomass of the culture increases in a characteristic fashion, until nutrients in the medium are exhausted and/or toxic by-products build up to inhibitory levels—this is called the 'stationary phase'. If cells are left in the stationary phase for too long, they will die and the culture will be lost. Therefore, cells should be transferred as they enter the stationary phase. It is therefore important that the batch growth-cycle parameters are determined for each cell-suspension culture.

Protoplasts

Protoplasts are plant cells with the cell wall removed. Protoplasts are most commonly isolated from either leaf mesophyll cells or cell suspensions, although other sources can be used to advantage. Two general approaches to removing the cell wall (a difficult task without damaging the protoplast) can be taken—mechanical or enzymatic isolation. Mechanical isolation, although possible, often results in low yields, poor quality and poor performance in culture due to substances released from damaged cells. Enzymatic isolation is usually carried out in a simple salt solution with a high osmoticum, plus the cell wall degrading enzymes.

It is usual to use a mix of both cellulase and pectinase enzymes, which must be of high quality and purity. Protoplasts are fragile and easily damaged, and therefore must be cultured carefully. Liquid medium is not agitated and a high osmotic potential is maintained, at least in the initial stages. The liquid medium must be shallow enough to allow aeration in the absence of agitation. Protoplasts can be plated out on to solid medium and callus produced. Whole plants can be regenerated by organogenesis or somatic embryogenesis from this callus. Protoplasts are ideal targets for transformation by a variety of means.

Root Cultures

Root cultures can be established *in vitro* from explants of the root tip of either primary or lateral roots and can be cultured on fairly simple media. The growth of roots *in vitro* is potentially unlimited, as roots are indeterminate organs. Although the establishment of root cultures was one of the first achievements of modern plant tissue culture, they are not widely used in plant transformation studies.

Shoot tip and Meristem Culture

The tips of shoots (which contain the shoot apical meristem) can be cultured *in vitro*, producing clumps of shoots from either axillary or adventitious buds. This method can be used for clonal propagation. Shoot meristem cultures are potential alternatives to the more commonly used methods for cereal regeneration as they are less genotype-dependent and more efficient (seedlings can be used as donor material).

Embryo Culture

Embryos can be used as explants to generate callus cultures or somatic embryos. Both immature and mature embryos can be used as explants.

Immature, embryo-derived embryogenic callus is the most popular method of monocot plant regeneration.

Microspore Culture

Haploid tissue can be cultured *in vitro* by using pollen or anthers as an explant. Pollen contains the male gametophyte, which is termed the 'microspore'. Both callus and embryos can be produced from pollen. Two main approaches can be taken to produce *in vitro* cultures from haploid tissue. The first method depends on using the anther as the explant. Anthers (somatic tissue that surrounds and contains the pollen) can be cultured on solid medium (agar should not be used to solidify the medium as it contains inhibitory substances). Pollen-derived embryos are subsequently produced via dehiscence of the mature anthers.

The dehiscence of the anther depends both on its isolation at the correct stage and on the correct culture conditions. In some species, the reliance on natural dehiscence can be circumvented by cutting the wall of the anther, although this does, of course, take a considerable amount of time. Anthers can also be cultured in liquid medium, and pollen released from the anthers can be induced to form embryos, although the efficiency of plant regeneration is often very low.

Immature pollen can also be extracted from developing anthers and cultured directly, although this is a very time-consuming process. Both methods have advantages and disadvantages. Some beneficial effects to the culture are observed when anthers are used as the explant material. There is, however, the danger that some of the embryos produced from anther culture will originate from the somatic anther tissue rather than the haploid microspore cells. If isolated pollen is used there is no danger of mixed embryo formation, but the efficiency is low and the process is time-consuming. In microspore culture, the condition of the donor plant is of critical importance, as is the timing of isolation.

Pretreatments, such as a cold treatment, are often found to increase the efficiency. These pretreatments can be applied before culture, or, in some species, after placing the anthers in culture. Plant species can be divided into two groups, depending on whether they require the addition of plant growth regulators to the medium for pollen/anther culture; those that do also often require organic supplements, *e.g.*, amino acids.

Many of the cereals (rice, wheat, barley and maize) require medium supplemented with plant growth regulators for pollen/anther culture. Regeneration from microspore explants can be obtained by direct embryogenesis, or via a callus stage and subsequent embryogenesis. Haploid tissue cultures can also be initiated from the female gametophyte (the ovule). In some cases, this is a more efficient method than using pollen or anthers. The ploidy of the plants obtained from haploid cultures may not be haploid.

This can be a consequence of chromosome doubling during the culture period. Chromosome doubling (which often has to be induced by treatment with chemicals such as colchicine) may be an advantage, as in many cases haploid plants are not the desired outcome of regeneration from haploid tissues. Such plants are often referred to as 'di-haploids', because they contain two copies of the same haploid genome.

PATHOGEN FREE PLANTS AND TISSUE CULTURE

Another purpose for which plant tissue culture is uniquely suited is in the obtaining, maintaining, and mass propagating of specific pathogen-free plants. The concept behind indexing plants free of pests is closely allied to the concept of using tissue culture as a selection system. Plant tissues known to be free of the pathogen under consideration (viral, bacterial, or fungal) are physically selected as the explant for tissue culture. In most cases, the apical domes of rapidly elongating shoot tips are chosen.

These are allowed to enlarge and proliferate under the sterile conditions of *in vitro* culture with the resulting plantlets tested for presence of the pathogen (a procedure called indexing). Cultures which reveal the presence of the pathogen are destroyed, while those which are indexed free of pathogen are maintained as a stock of pathogen-free material. Procedures similar to these have been used successfully to obtain virus-free plants of a number of species and bacteria-free plants of species known to have certain leaf spot diseases. The impact of obtaining pathogen-free nursery stock can only be speculative, since little research documenting viral, bacterial, or fungal diseases transmitted through propagation of woody ornamentals is available.

SOMATIC HYBIDISATION

The ability to fuse plant cells from species which may be incompatible as sexual crosses and the ability of plant cells to take up and incorporate foreign genetic codes extend the realm of plant modifications through tissue culture to the limits of the imagination. Most such manipulations are carried out using plant "protoplasts". Protoplasts are single cells which have been stripped of their cell walls by enzymatic treatment. A single leaf treated under these conditions may yield tens of millions of single cells, each theoretically capable of eventually producing a whole plant. This concept has fueled speculation as diverse as the possibilities of obtaining nitrogen-fixing corn plants on the one extreme to discovering a yellow-flowered African violet on the other extreme.

The observation that has provided the impetus for most of this research is that when cells are stripped of their cell walls and brought into close contact, they tend to fuse with each other. This "somatic hybridisation" is not subject to the same incompatibility problems that limit traditional plant breeding strategies. It is conceivable then that one could hybridise a Juneberry with a crabapple or a plum, but the fundamental research required to demonstrate such an event has yet to be conducted. The potential use of somatic hybridisation to bring about novel combinations of genetic material has been demonstrated in the genera Petunia and Nicotiana. Research funded in part by the Horticultural Research Institute at the University of Wisconsin is investigating the feasibility of using such techniques with woody species. Brent McGown and co-workers have succeeded in obtaining naked cells from tissue cultures of Betula and Rhododendron, but as of yet, they have neither obtained plants from single cells not achieved cellular fusion.

However, further research in this area promises to have a tremendous impact on our concepts of woody plant diversity. Just as remarkable as the idea of fusing plant protoplasts is the idea of incorporating foreign genetic material into the genetic code of plant cells. Such transformations have been carried out in the so-called "gene-splicing" experiments where the information for making insulin was incorporated into bacteria. Not only is the desired information transmitted to succeeding generations of bacteria, but the bacterial cultures become synthesisers of insulin as well. Plant cells can be made to take up foreign genetic codes, but evidence that this can be transmitted into the daughter cells and serve the intended function is lacking.

Plant tissue culture research is multi-dimensional. While most nurserymen have been introduced to the techniques and advantages of micropropagation, few have ventured to use it as a propagation tool. The applicability of micropropagation for woody trees has been demonstrated as feasible since all aspects of the technology have confirmed the fact that trees produced by this method look like and grow like their counterparts produced by traditional methods of cloning.Other dimensions of tissue culture research have been less well publicised.

The potential for selecting pathogen free plants, for selecting stress-tolerant and pathogen-resistant clones of plants, and the novel genetic combinations to be achieved through somatic hybridisation are all lines of research which can have a profound impact on the nursery industry.

UTILISING ANTHER CULTURE TO SELECT FOR COLD HARDINESS

Crosses between distantly related species can bring together novel gene combinations. However, the hybrid offspring can be few in number, genetically unstable and require years of further selection and screening before any advantageous characteristics can be brought near to commercial use. Anther culture (androgenesis), to generate haploid plants from pollen microspores, is one way to shorten this process. It allows novel allele combinations, particularly ones involving recessive characters, to be assessed in intact plants. Useful individuals can then be developed into homozygous and fertile plants through chromosome doubling techniques, and brought into a breeding programme.We have recently been involved in a collaborative project with the Institute of Grassland and Environmental Research (IGER) to use this approach to improve cold-tolerance and fodder quality in grasing grasses.

Crosses between *Lolium multiflorum* (Italian ryegrass) and *Festuca arundinacea* (tall fescue) should offer valuable combinations of characteristics. The *Lolium* species should provide good growth characteristics, while the *Festuca* provides stress-tolerance. One hybrid individual (*Festulolium*) resulting such a cross had already shown drought-tolerance characteristics. However, the out-breeding nature of these grass species, along with the hexaploid genome of *F. arundinacea* and autotetraploid *L. multiflorum* indicated that a lengthy breeding programme might be necessary.The research project therefore aimed to produce androgenic plants from the existing pentaploid *Festulolium* plant and assess them for cold tolerance.

This exploited the expertise in tissue culture at the University of Liverpool together with experience in breeding for stress-tolerance at IGER.Anthers containing immature pollen (microspores) are the starting material for androgenesis. Flowers have to be selected at the correct developmental stage, which varies from species to species. In addition, some individual genotypes may not be amenable to anther culture, or require specific pretreatments. Careful microscopy and testing of successful pre-treatments of related species are therefore necessary when dealing with a new species. For the Graminae, microspores must be at the mononucleate stage and no pre-treatment is necessary. The cut flowers were surface sterilised and opened in sterile conditions under a binocular microscope.

The anthers were dissected and transferred to a solid nutrient medium. Large numbers could be placed on each petri dish. Callus developed, which was transferred to a different medium to initiate embryos and so generate haploid plants. Over 200 androgenic plants were produced at Liverpool, each originating from a different microspore. Each therefore represented a genetically different individual. Testing for phosphoglucosisomerase, where a different isozyme was contributed by each of the five chromosome groups within the *Festulolium* plant, indicated that the pollen-derived plants had a wide variety of chromosome combinations from each of the parents of the hybrid.

The freesing-tolerance of these plants varied considerably, with three individuals able to survive the extreme cold of -14 degrees Celsius. When the chromosome complement of two of these plants was examined using genomic *in situ* hybridisation (GISH), they carried virtually the whole genome of *F. pratensis*, a parent of *F. arundinacea* noted for its freesing-tolerance. Unfortunately, the fertility of these two plants was not restored by chromosome doubling, so that they could not be used for further breeding. However, they demonstrated the potential of androgenesis for rapid assessment of the genetic potential available from a difficult breeding combination, indicating that this type of wide cross revealed characters of cold and drought tolerance that were worth pursuing.

Using Somaclonal Variation to Select for Disease Resistance

Plant tissue cultures isolated from even a single cell can show variation after repeated subculture. Distinct lines can be selected with their own particular morphology and physiology. It suggests that the tissue culture contains a population of genotypes whose proportion can be altered by imposing an appropriate selection pressure. This variation can be transmitted to plants regenerated from the tissue cultures, and is called somaclonal variation. It provides an additional source of novel variation for exploitation by plant breeders.

The carrot cultivar Fancy was used in our laboratory to generate a series of 197 regenerant progeny lines. These plants showed considerable morphological variation. They were tested for resistance to the leaf spot pathogen *Alternaria dauci,* which can cause total necrosis of mature leaves. They had a greater degree of variation in response than the parental cultivar.

DEMONSTRATION OF TISSUE CULTURE FOR TEACHING

The starting point for all tissue cultures is plant tissue, called an explant. It can be initiated from any part of a plant - root, stem, petiole, leaf or flower - although the success of any one of these varies between species. It is essential that the surface of the explant is sterilised to remove all microbial contamination. Plant cell division is slow compared to the growth of bacteria and fungi, and even minor contaminants will easily over-grow the plant tissue culture. The explant is then incubated on a sterile nutrient medium to initiate the tissue culture. The composition of the growth medium is designed to both sustain the plant cells, encourage cell division, and control development of either an undifferentiated cell mass, or particular plant organs.

The concentration of the growth regulators in the medium, namely auxin and cytokinin, seems to be the critical factor for determining whether a tissue culture is initiated, and how it subsequently develops. The explant should initially form a callus, from which it is possible to generate multiple embryos and then shoots, forming the basis for plant regeneration and thus the technology of micropropagation. The first stage of tissue culture initiation is vital for information on what combination of media components will give a friable, fast-growing callus, or a green chlorophyllous callus, or embryo, root or shoot formation. There is at present no way to predict the exact growth medium, and growth protocol, to generate a particular type of callus. These characteristics have to be determined through a carefully designed and observed experiment for each new plant species, and frequently also for each new variety of the species which is taken into tissue culture. The basis of the experiment will be media and protocols that give the desired effect in other plant species, and experience.

The Demonstration

The strategy for designing a medium to initiate tissue culture, showing how growth regulators and other factors modulate development, can be demonstrated using the African Violet, a popular house plant. Leaf sections are the source of explants. This demonstration is regularly carried out by a student class, and gives reliable results.

Sterile supplies are provided from central facilities, and provision of sterile working areas (for example, in laminar flow hoods) is an advantage, although cultures can be initiated in an open laboratory with careful aseptic technique. The standard precautions used during any laboratory work involving chemicals or microbes should be adopted. If you are in any doubt about safety hazards associated with this demonstration, you should consult your local safety adviser.

Step 1: Selection of the Leaves

Leaves are cut from healthy plants, leaving a short length of petiole attached. They should be selected to each yield several explants of leaf squares with approximately 1 cm sides. The youngest and oldest leaves should be avoided. Wash the dust off the leaves in a beaker of distilled water, holding the leaf stalk with forceps.

Step 2: Surface Sterilisation and Preparation of the Explants

This part of the procedure should be carried out in a sterile working area, or with meticulous aseptic technique. The leaf, with the petiole still attached, should be immersed in 70 per cent ethanol for 30 seconds, then transferred to a sterile petri dish. Sterile scissors and forceps are then used to cut squares from the leaf as explants, each with approximately 1 cm sides. The explants are transferred into a 10 per cent hypochlorite bleach solution for 5 minutes, gently agitating once or twice during this time. They are then washed free of bleach by immersing in four successive beakers of sterile distilled water, leaving them for 2-3 minutes in each. Three explants are placed on each petri dish of growth medium, with the upper epidermis pressed gently against the surface of the agar to make good contact. The petri dishes are sealed with plastic film to prevent moisture loss, and incubated at 25°C in 16h light/8h dark.

Step 3: Assessment of Tissue Culture Development

The explants are incubated for 4 - 6 weeks, and inspected at weekly or fortnightly intervals. The growth of obvious bacterial or fungal colonies indicates contamination, and data from such cultures is obviously suspect. The development of dark brown tissue cultures can also be a consequence of contamination.The media used in the demonstration are designed to show the effects of auxin, cytokinin, sucrose and mineral salts on development. The media were based on the well-known Murashige and Skoog inorganic medium.

Typical Results

Absence of sucrose inhibits this production. Shoot production is also limited on the low sucrose concentration, but comparable with the control at high sucrose.At zero and low levels of cytokinin, callus forms where the leaf surface is in contact with the medium, while at high levels, shoot formation is stimulated.At zero and low levels of auxins there is a stimulus to shoot formation, but at high concentrations, large numbers of roots are formed.At low and zero levels of MS salts, there is no growth at all.These very obvious variations demonstrate the importance of a carbon and inorganic salt source for plant growth, as well as the effect of the auxin:cytokinin ration on the control of plant development.

PLANT IMPROVEMENT THROUGH TISSUE CULTURE

In introducing this research update, it was mentioned that the major impact of tissue culture technology would not be in the area of micropropagation, but rather in the area of controlled manipulations of plant germplasm at the cellular level. The ability to unorganise, rearrange, and reorganise the constituents of higher plants has been demonstrated with a few model systems to date, but such basic research is already being conducted on ornamental trees and shrubs with the intent of obtaining new and better landscape plants.

SELECTION OF PLANTS WITH ENHANCED STRESS OR PEST RESISTANCE

Perhaps the most heavily researched area of tissue culture today is the concept of selecting disease, insect, or stress resistant plants through tissue culture. Just as significant gains in the adaptability of many species have been obtained by selecting and propagating superior individuals, so the search for these superior individuals can be tremendously accelerated using *in vitro* systems. Such systems can attempt to exploit the natural variability known to occur in plants or variability can be induced by chemical or physical agents known to cause mutations.

All who are familiar with bud sports, variegated foliage and other types of chimeras have an appreciation for the natural variability in the genetic makeup or expression in plants. Chimeras are the altered cellular expressions which are visible, but for each of these which are observed many more differences probably exist but are masked by the overall organisation of the plant as a whole. For example, even in frost-tender species, certain cells or groups of cells may be frost hardy. However, because most of the organism is killed by frost, the tolerant cells eventually die because they are unable to support themselves without the remainder of the organised plant. Plant tissues grown *in vitro* can be released from the organisation of the whole plant through callus formation. If these groups of cells are then subjected to a selection agent such as freesing, then those tolerant ones can survive while all those which are susceptible will be killed. This concept can be applied to many types of stress as well as resistance to fungal and bacterial pathogens and various types of phytotoxic chemical agents.

The goal of selecting such resistant cell lines would be to reorganise whole plants from them which would retain the selected resistance Current research in this area extends across many interests including attempts to select salt tolerant lines of tomato, freesing resistant tobacco plants, herbicide resistant agronomic crops, and various species of plants with enhanced pathogen resistance. Imagine, if you will, the impact of a fireblight-resistant Bartlett pear, a clone of pin oak for alkaline soils, or a selection of southern magnolia hardy to zone 4!

5

Techniques of Plant Tissue Culture

HISTORY

Plant tissue culture is a method or technique to isolate parts of plants (protoplasm, cells, tissues, and organs) and grow them on artificial media in aseptic conditions in a controlled space so that parts of these plants can grow and develop into complete plants. Plant tissue culture techniques are essential to many types of academic inquiry, as well as to many applied aspects of plant science. In the past, plant tissue culture techniques have been used in academic investigations of totipotency and the roles of hormones in cytodifferentiation and organogenesis. Currently, tissue-cultured plants that have been genetically engineered provide insight into plant molecular biology and gene regulation. Plant tissue culture techniques are also central to innovative areas of applied plant science, including plant biotechnology and agriculture. For example, select plants can be cloned and cultured as suspended cells from which plant products can be harvested. In addition, the management of genetically engineered cells to form transgenic whole plants requires tissue culture procedures; tissue culture methods are also required in the formation of somatic haploid embryos from which homozygous plants can be generated. Thus, tissue culture techniques have been, and still are, prominent in academic and applied plant science.

The techniques demonstrated in these exercises range from simple ones that can easily be performed by beginning students to those done by botany or physiology students. The use of tissue culture techniques in the beginning just to prove the theory of "totipotensi" ("total genetic potential") is expressed by Schleiden and Schwann which states that the plant cell as the smallest unit can grow and thrive if kept in appropriate conditions. We have used tissue culture techniques not only as a means for studying aspects of plant physiology and biochemistry, but has developed into methods for various purposes.

ASEPTIC TECHNIQUE

The essence of aseptic technique is the exclusion of invading microorganisms during experimental procedures. If sterile tissues are available, then the exclusion of microorganisms is accomplished by using sterile instruments and culture media concurrently with standard bacteriological transfer procedures to avoid extraneous contamination. Media and apparatus are rendered sterile by autoclaving at 15 lbs/inch 2 (121°C) for 15 minutes. The use of disposable sterile plasticware reduces the need for some autoclaving. Alternative sterilization techniques such as filter sterilization must be employed for heat-labile substances like cytokinins. Aseptic transfers can be made on the laboratory bench top by using standard bacteriological techniques (*i.e.*, flaming instruments prior to use and flaming the opening of receiving vessels prior to transfer). Aseptic transfers are more easily performed in a transfer chamber such as a laminar flow hood, which is also preferably equipped with a bunsen burner.

If experimental tissues are not aseptic, then surface sterilization procedures specific to the tissues are employed. Common sterilants are ethyl alcohol and/or chlorox with an added surfactant. Concentration of sterilants and exposure time are determined empirically. When cultured in vitro, all the needs of the plant cells, both chemical and physical, have to met by the culture vessel, the growth medium, and the external environment (light, temperature, etc.).

The growth medium has to supply all the essential mineral ions required for growth and development. In many cases (as the biosynthetic capability of cells cultured in vitro may not replicate that of the parent plant), it must also supply additional organic supplements such as amino acids and vitamins. Many plant cell cultures, as they are not photosynthetic, also require the addition of a fixed carbon source in the form of a sugar (most often sucrose). One other vital component that must also be supplied is water, the principal biological solvent. Physical factors, such as temperature, pH, the gaseous environment, light (quality and duration), and osmotic pressure, also have to be maintained within acceptable limits.

PLANT TISSUES

Plants are composed of three major organ groups: roots, stems and leaves. As we know from other areas of biology, these organs are comprised of tissues working together for a common goal (function). In turn, tissues are made of a number of cells which are made of elements and atoms on the most fundamental level. In this section, as suggested look at the various types of plant tissue and their place and purpose within a plant. It is important to realise that there may be slight variations and modifications to the basic tissue types in special plants.

Plant tissues are characterized and classified just as to their structure and function. The organs that they form will be organized into patterns within a plant which will aid in further classifying the plant. A good example of this is the three basic tissue patterns found in roots and stems which serve to delineate between woody dicot, herbaceous dicot and monocot plants. As suggested look at these classifications later on in the tutorial.

MERISTEMATIC TISSUES

Tissues where cells are constantly dividing are called meristems or meristematic tissues. These regions produce new cells. These new cells are generally small, six-sided boxlike structures with a number of tiny vacuoles and a large nucleus, by comparison. Sometimes there are no vacuoles at all. As the cells mature the vacuoles will grow to many different shapes and sizes, depending on the needs of the cell. It is possible that the vacuole may fill 95 per cent or more of the cell's total volume.

There are three types of meristems:

- Apical Meristems
- Lateral Meristems
- Intercalary Meristems

Apical meristems are located at or near the tips of roots and shoots. As new cells form in the meristems, the roots and shoots will increase in length. This vertical growth is also known as primary growth. A good example would be the growth of a tree in height. Each apical meristem will produce embryo leaves and buds as well as three types of primary meristems: protoderm, ground meristems, and procambium. These primary meristems will produce the cells that will form the primary tissues.

Lateral meristems account for secondary growth in plants. Secondary growth is generally horizontal growth. A good example would be the growth of a tree trunk in girth. There are two types of lateral meristems to be aware of in the study of plants. The vascular cambium, the first type of lateral meristem, is sometimes just called the cambium. The cambium is a thin, branching cylinder that, except for the tips where the apical meristems are located, runs the length of the roots and stems of most perennial plants and many herbaceous annuals. The cambium is responsible for the production of cells and tissues that increase the thickness, or girth, of the plant. The cork cambium, the second type of lateral meristem, is much like the vascular cambium in that it is also a thin cylinder that runs the length of roots and stems. The difference is that it is only found in woody plants, as it will produce the outer bark.

Both the vascular cambium and the cork cambium, if present, will begin to produce cells and tissues only after the primary tissues produced by the apical meristems have begun to mature. Intercalary meristems are found in grasses and related plants that do not have a vascular cambium or a cork cambium, as they do not increase in girth. These plants do have apical meristems and in areas of leaf attachment, called nodes, they have the third type of meristematic tissue. This meristem will also actively produce new cells and is responsibly for increases in length. The intercalary meristem is responsible for the regrowth of cut grass. There are other tissues in plants that do not actively produce new cells. These tissues are called non-meristematic tissues. Non-meristematic tissues are made of cells that are produced by the meristems and are formed to various shapes and sizes depending on their intended function in the plant. Sometimes the tissues are composed of the same type of cells throughout, or sometimes they are mixed. There are simple tissues and complex tissues to consider, but as suggested start with the simple tissues for the sake of discussion.

Simple Tissues

There are three basic types, named for the type of cell that makes up their composition.

- Parenchyma cells form parenchyma tissue. Parenchyma cells are the most abundant of cell types and are found in almost all major parts of higher plants. These cells are basically sphere shaped when they are first made. However, these cells have thin walls, which flatten at the points of contact when many cells are packed together. Generally, they have many sides with the majority having 14 sides. These cells have large vacuoles and may contain various secretions including starch, oils, tannins, and crystals. Some parenchyma cells have many chloroplasts and form the tissues found in leaves. This type of tissue is called chlorenchyma. The chief function of this type of tissue is photosynthesis, while parenchyma tissues without chloroplasts are generally used for food or water storage. Additionally, some groups of cells are loosely packed together with connected air spaces, such as in water lilies, this tissue is called aerenchyma tissue. These type of cells can also develop irregular extensions of the inner wall which increases overall surface area of the plasma membrane and facilitates transferring of dissolved substances between adjacent cells. Parenchyma cells can divide if they are mature, and this is vital in repairing damage to plant tissues. Parenchyma cells and tissues comprise most of the edible portions of fruit.
- Collenchyma cells form collenchyma tissue. These cells have a living protoplasm, like parenchyma cells, and may also stay alive for a long period of time. Their main distinguishing difference from parenchyma cells is the increased thickness of their walls. In cross section, the walls looks uneven. Collenchyma cells are found just beneath the epidermis and generally they are elongated and their walls are pliable in addition to being strong. As a plant grows these cells and the tissues they form, provide flexible support for organs such as leaves and flower parts. Good examples of collenchyma plant cells are the 'strings' from celery that get stuck in our teeth.

- Sclerenchyma cells form sclerenchyma tissue. These cells have thick, tough secondary walls that are imbedded with lignin. At maturity, most sclerenchyma cells are dead and function in structure and support.

  Sclerenchyma cells can occur in two forms:

  - Sclereids are sclerenchyma cells that are randomly distributed throughout other tissues. Sometimes they are grouped within other tissues in specific zones or regions. They are generally as long as they are wide. An example, would be the gritty texture in some types of pears. The grittiness is due to groups of sclereid cells. Sclereids are sometimes called stone cells.
  - Fibres are sometimes found in association with a wide variety of tissues in roots, stems, leaves and fruits. Usually fibre cells are much longer than they are wide and have a very tiny cavity in the centre of the cell. Currently, fibres from over 40 different plant families are used in the manufacture of textiles, ropes, string and canvas goods to name a few.

Secretory Cells and Tissues

As a result of cellular processes, substances that are left to accumulate within the cell can sometimes damage the protoplasm. Thus it is essential that these materials are either isolated from the protoplasm in which they originate, or be moved outside the plant body. Although most of these substances are waste products, some substances are vital to normal plant functions. Examples: oils in citrus, pine resin, latex, opium, nectar, perfumes and plant hormones. Generally, secretory cells are derived from parenchyma cells and may function on their own or as a tissue. They sometimes have great commercial value.

Complex Tissues

Tissues composed of more than one cell type are generically referred to as complex tissues. Xylem and phloem are the two most important complex tissues in a plant, as their primary functions include the transport of water, ions and soluble food substances throughout the plant. While some complex tissues are produced by apical meristems, most in woody plants are produced by the vascular cambium and is often referenced as vascular tissue. Other complex tissues include the epidermis and the periderm.

The epidermis consists primarily of parenchyma-like cells and forms a protective covering for all plant organs. The epidermis includes specialized cells that allow for the movement of water and gases in and out of the plant, secretory glands, various hairs, cells in which crystals are accumulated and isolated, and other cells that increase absorption in the roots. The periderm is mostly cork cells and therefore forms the outer bark of woody plants. It is considered to be a complex tissue because of the pockets of parenchyma cells scattered throughout.

Xylem

Xylem is an important plant tissue as it is part of the 'plumbing' of a plant. Think of bundles of pipes running along the main axis of stems and roots. It carries water and dissolved substances throughout and consists of a combination of parenchyma cells, fibres, vessels, tracheids and ray cells. Long tubes made up of individual cells are the vessels, while vessel members are open at each end. Internally, there may be bars of wall material extending across the open space. These cells are joined end to end to form long tubes. Vessel members and tracheids are dead at maturity.

Tracheids have thick secondary cell walls and are tapered at the ends. They do not have end openings such as the vessels. The tracheids ends overlap with each other, with pairs of pits present. The pit pairs allow water to pass from cell to cell. While most conduction in the xylem is up and down, there is some side-to-side or lateral conduction via rays. Rays are horizontal rows of long-living parenchyma cells that arise out of the vascular cambium. In trees, and other woody plants, ray will radiate out from the centre of stems and roots and in cross-section will look like the spokes of a wheel.

Phloem

Phloem is an equally important plant tissue as it also is part of the'plumbing' of a plant. Primarily, phloem carries dissolved food substances throughout the plant. This conduction system is composed of sieve-tube member and companion cells, that are without secondary walls. The parent cells of the vascular cambium produce both xylem and phloem. This usually also includes fibres, parenchyma and ray cells. Sieve tubes are formed from sieve-tube members laid end to end.

The end walls, unlike vessel members in xylem, do not have openings. The end walls, however, are full of small pores where cytoplasm extends from cell to cell. These porous connections are called sieve plates. In spite of the fact that their cytoplasm is actively involved in the conduction of food materials, sieve-tube members do not have nuclei at maturity. It is the companion cells that are nestled between sieve-tube members that function in some manner bringing about the conduction of food. Sieve-tube members that are alive contain a polymer called callose.

Callose stays in solution as long at the cell contents are under pressure. As a repair mechanism, if an insect injures a cell and the pressure drops, the callose will precipitate. However, the callose and a phloem protein will be moved through the nearest sieve plate where they will for a plug. This prevents further leakage of sieve tube contents and the injury is not necessarily fatal to overall plant turgor pressure.

Epidermis

The epidermis is also a complex plant tissue, and an interesting one at that. Officially, the epidermis is the outermost layer of cells on all plant organs. The epidermis is in direct contact with the environment and therefore is subject to environmental conditions and constraints. Generally, the epidermis is one cell layer thick, however there are exceptions such as tropical plants where the layer may be several cells thick and thus acts as a sponge. Cutin, a fatty substance secreted by most epidermal cells, forms a waxy protective layer called the cuticle.

The thickness of the cuticle is one of the main determiners of how much water is lost by evaporation. Additionally, at no extra charge, the cuticle provides some resistance to bacteria and other disease organisms. Some plants, such as the wax palm, produce enough cuticle to have commercial value: carnauba wax. Other wax products are used as polishes, candles and even phonographic records. Epidermal cells are important for increasing absorptive surface area in root hairs. Root hairs are essentially tubular extensions of the main root body composed entirely of epidermal cells. Leaves are not left out. They have many small pores called stomata that are surrounded by pairs of specialized epidermal cells called guard cells. Guard cells are unique epidermal cells because they are of a different shape and contain chloroplasts. They will be discussed in detail later on in the tutorial. There are other modified epidermal cells that may be glands or hairs that repel insects or reduce water loss.

Periderm

In woody plants, when the cork cambium begins to produce new tissues to increase the girth of the stem or root the epidermis is sloughed off and replaced by a periderm. The periderm is made of semi-rectangular and boxlike cork cells. This will be the outermost layer of bark. These cells are dead at maturity. However, before the cells die, the protoplasm secretes a fatty substance called suberin into the cell walls. Suberin makes the cork cells waterproof and aids in protecting tissues beneath the bark.

There are parts of the cork cambium that produce pockets of loosely packed cork cells. These cork cells do not have suberin imbedded in their cell walls. These loose areas are extended through the surface of the periderm and are called lenticels. Lenticels function in gas exchange between the air and the stem interior. At the bottom of the deep fissures in tree bark are the lenticels.

TISSUE GROWTH

Analogous considerations, as applied in the foregoing to cell growth, apply to the growth of tissues and organs. Just as the molecular population of the cell consists of a reproductive and a nonreproductive fraction, so the cellular population of a tissue is composed of proliferative and nonproliferating components. In most organs proliferation becomes gradually confined to certain restricted areas, so-called germinal zones, arranged in layers (*e.g.* central nervous system, skin), cords (*e.g.* sex cords) or foci (*e.g.* buds). Of the newly formed cells, some remain in the germinal zone, while the rest move away from it. The former continue to proliferate, while the latter cease to multiply and usually undergo some terminal specialization.

This dichotomy has sometimes been interpreted as signifying an antagonistic relation between proliferation and differentiation.9 Statistically speaking, it is correct that as differentiation progresses, proliferation declines. But this does not imply that differentiation in the strict sense and proliteration are mutually exclusive. From our earlier discussion, it should be plain that the proliferating sources of epidermis, bone, blood, muscle, glands, etc. are differentiated to the extent that each can reproduce only its own kind and none of the others. Yet they are still actively multiplying.

Only the full transformation into terminal stages seems to be held up in those cells that are kept in mitotic agitation. Proliferation does not interfere with differentiation, but it does impede the elaboration of certain manifest products of differentiation. Conversely, the general impairment of proliferative capacity in cells undergoing terminal specialization can be ascribed to the fact that most of their substance, including mitotic prerequisites, is diverted into the building of differentiation products (*e.g.* intracellular fibre systems).

In early developmental stages practically all cells proliferate. With progressive differentiation, however, the outlined segregation into reproductive and sterile groups takes place. Since the latter do not further contribute to the increase in mass, the total growth of a tissue or organ will vary directly with the ratio of reproductive to nonreproductive cells. This ratio in turn depends on the geometric configuration of the proliferating zone and the rate at which newly proliferated cells desert it.

Further variability is introduced by the fact that most organs receive secondary additions from foreign sources (*e.g.* blood vessels, nerves, pigment cells), and, conversely, lose some of their own cells by dispersion or destruction. Since each organ has its own rule, it is evident that bulk measurements, which fail to separate the actual sources of the measured materials from the "dead weight," can furnish no valid basis for comparing "differential" growth of parts differing in constitution. If percentage increase is faster in one organ than in another, this could be due to either:

- A larger initial proportion of reproductive cells,
- A larger quota of retention of new cells in the germinal zone,
- Greater infiltration of foreign elements,
- Less dissipation of cells and cell products,
- An intrinsically higher rate of reproduction in the respective germinal cells.

Only differences not accounted for by are to be ascribed to that is, to a higher "growth rate." For an illustration let us choose the growing vertebrate eye. Due to its simple form and accessibility it has been a favourite object for comparative studies on growth, with the increase in dimensions serving as index. How heterogeneous a collection of events this index covers can be readily seen from the following account. The original eye vesicle consists of a certain initial allotment of cells from the embryonic brain wall. At first, all of these cells divide.

The growth function at this stage is therefore a volume function. In the cup stage the retina becomes multilayered, with a sharp division into a germinal and a sterile zone. Only the cell layer in contact with the outer surface, corresponding to the ventricular (ependymal) layer of the brain, continues to proliferate, while the cells released into deeper layers differentiate the various retinal strata without further multiplication. The source of growth thus has become reduced to a two-dimensional one, causing a marked decline in the relative growth rate taken over the whole organ (*e.g.* from measurements of diameter). Later, the cells of the germinal layer themselves cease to proliferate and transform into sensory cells, a process which starts from the centre (macula) and spreads rapidly toward the periphery (ciliary zone) of the retina. Eventually, only the cells at the rim retain residual capacity to multiply.

Further growth is then essentially by apposition from this rim; that is, the growth source has shrunk from planar to linear extension. Meanwhile some of the neuroblasts, though no longer multiplying, grow in size as they sprout nerve processes, which, grouped into plexiform layers, add to the thickness of the retina. During the later stages a gelatinous secretion, supposed to come from cells of both retina and lens, fills the interior with vitreous humour, thereby progressively distending the eyeball. In addition, blood vessels and other mesenchym penetrate into the eye from the surroundings. This diversity and complexity of the component processes contributing to eye size makes the search for a single "growth-controlling" principle appear utterly unrealistic, and comparisons between different eyes on the sole basis of size must be fallacious.

First, what causes the confinement of proliferation to the single cell layer at the convexity of the retina? The answer is unknown. But a clue lies in the fact that in this and most similar instances the site of proliferation corresponds to some prominent geometrical feature of the system, such as a surface or fold or tip or crest. For the cells concerned this means unique physical conditions (*e.g.* tensions, pressures, unique exposure to surrounding agencies) not shared by the rest of the tissue. In neural tube and retina, for instance, the direct exposure to the fluid of the central canal and ventricular spaces might be the crucial factor. The fact that injury can activate cells of the interior to take part in proliferation could be explained by the penetration of the central fluid to the deeper layers as a result of the break in the internal limiting membrane. This is purely conjectural, but capable of an experimental test. Next, what determines the further fate of the cells proliferated in the germinal layer? Of the two daughter cells of a germinal cell, either:

- Both stay in the germinal layer
- Both move away
- One stays and the other moves off. The same alternatives recur at subsequent divisions. Continuation of procedure
- Would lead to exponential increase of the proliferative source with no cells left for terminal differentiation, while procedure
- Would rapidly deplete the proliferative source. The actual growth of the eye at this phase is therefore determined by the relative incidence of events.

But again, factual information on this point is practically wholly lacking. In anuran amphibians the number of dividing cells in the retina has been reported to remain remarkably constant, which would mean that all divisions follow scheme.

But this is definitely not true of the chick, for instance, in which the number of proliferating cells increases with the expansion of the retina. Here the rate of growth is therefore largely dependent on the statistical apportionment of the newly formed cells according to:

- This in turn seems to resolve itself into a matter of the relative orientation assumed by the daughter cells, either during mitosis or immediately after, tangential orientation leading
- Radial orientation
- A plausible hypothesis is the following.

It has been demonstrated that dividing cells separate in the direction of maximum cell elongation, which corresponds to the direction of maximum tension. Hertwig's rules of cleavage are an illustration of this principle. Assuming that retinal cells follow the same principle, we may conclude that distribution of type:

- Will result whenever the locally prevailing tension is tangential, while types
- Will result from radial tension or lateral pressure. According to this concept the proliferating layer can expand only as long as the retina is under tangential tension. The source of such force is found in the distension of the eyeball by the growing lens and the vitreous body. Retina, lens, and vitreous body would thereby form a self-regulating system operating as follows.

Accumulation of vitreous fluid would stretch the retina, producing a prevalence of tangential tensions, which, in turn, would cause a large proportion of newly proliferated cells to stay in the germinal surface according to:

- As these additional cells are intercalated, the lateral stretch is reduced and may eventually even be reversed into pressure if the cells begin to crowd laterally; in that case, radial movement

- Meanwhile, however, further secretion of vitreous humour has produced further distension so that more new cells can be accommodated in the surface expanse, and the cycle repeats itself. Because of widely varying local conditions, these events will not cause a regular alternation of phases of tangential and radial divisions as in cleavage, but will merely statistically increase the probability of the occurrence of one or the other. There are some experimental results suggesting such a mechanism of growth regulation. When the vitreous humour of an embryonic chick eye is drained, the collapsed retina remains greatly reduced in size, but increases in thickness. The loss of a progressively distending core would properly account for the result.

It has also been demonstrated that when incongruous eye-lens combinations are produced by heteroplastic transplantation, a mutual adjustment of size occurs. This could be ascribed to the fact that an undersized lens, both because of its smallness and because it will permit vitreous body to flow out, will not be able to exercise the distensive function which promotes retinal growth until it has caught up with the size of the eye and sealed the borders. Reciprocally, an oversized lens will produce greater distension, hence, expansion of the germinal layer, resulting in intensified retinal growth.

In a similar capacity, distension of vesicles, glands, ducts, etc. by their own or foreign discharges may be a major factor in regulating the extent of their proliferating layers, and consequently of the amount of growth. Perhaps the excessive size of the gut in amphibians fed on a bulky vegetable diet as compared with meat-fed ones is due to the same principle. One could even envisage that the air in the larval lungs of gill-breathing aquatic amphibians is there for a growth-supporting rather than respiratory purpose.

These examples may suffice to illustrate the dependence of overall "growth rate" on the special mode of growth of each organ, particularly the geometric configuration of its germinal zone.

ORIENTATION OF GROWTH

Growth as such, *i.e.* the mere increase of the specific molecular population of the cell, is a scalar process. It has no intrinsic direction. Its directiveness is given to it by vector components of the physical frame-work in which it occurs. Yet growth itself plays a decisive role in creating those physical conditions which, in turn, will orient its further course. A few examples may illustrate this principle of reverberation.

In the preceding section reference was made to the effect of tension on the form of growth in epithelial tissues. Even more spectacular is its effect on the architecture of tissues of mesenchymal origin. Connective tissue, tendon, fascia, cartilage, bone, blood vessels, muscles, all respond to mechanical tensions by orientation of the component cells and intercellular fibre systems along the lines of stress. Experimental analysis of this response in tissue culture showed that the observed cell orientation is a secondary product following a primary orientation of the colloidal matrix in which the cells lie embedded. The sequence is as follows:

- Tension produces alignment of aggregates of the anisodiametric molecules of the matrix into chains.
- Such chains join to build up submicroscopic fibrils, which eventually may merge into microscopic fibres.
- Cytoplasmic filaments (filopodia) from cells inhabiting the matrix extend along the interfaces between these solid micellar threads and the interstitial liquid.
- The rest of the cell body follows the advancing filopodia.

In this manner the cell becomes greatly elongated and is made to trace the ultrastructural pattern of its surroundings. Nerve processes advance in the same fashion, except that their cell body remains anchored. In the original tissue culture experiments the matrix consisted of blood plasma, with fibrin constituting the orientable ultrastructural reticulum.

Later experiments revealed that even in so-called cultures in liquid medium, cell orientation is brought about by a similar reticulum, namely, coagulable fibrous material exuding from the cells themselves, spreading along the surfaces of the liquid and constituting a trellis ("ground mat") on which the cells then advance.

Tensional fibrin organization in blood clots has likewise been disclosed as the mechanism orienting cell growth in wound healing and nerve regeneration in the body. In embryonic tissues, other fibre proteins can be expected to operate in like manner. All pertinent observations and experiments made thus far confirm the conclusion that tension, by way of its organizing effect on the colloidal cellular milieu, can determine:

- Orientation of cell movement,
- Cell elongation,
- Cell shape,
- Orientation of mitotic spindles, hence, direction of division,
- Rate of movement,
- Rate of multiplication,
- Deposition of fibrous cell products.

Figure illustrates diagrammatically how different degrees of stretch applied to a fibrous colloid translate themselves into the morphology of inhabitant, as well as immigrating, cells. Potent as these effects are, they could not be a major factor in development if only tensions of extraneous origin were involved.

But tensional stresses also arise within the developing organism as a result of shifts, growth, and chemical activity of its parts.

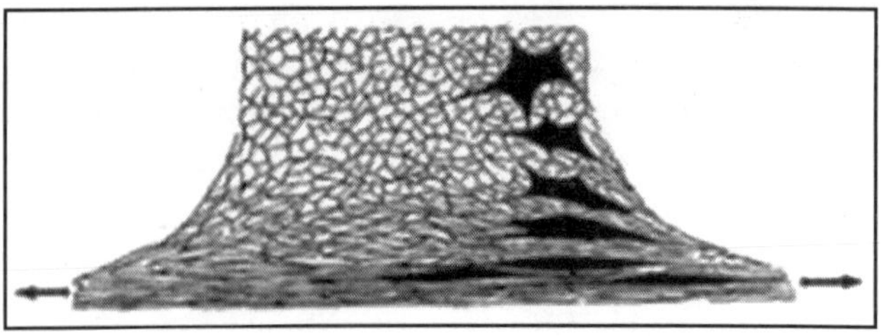

Fig. Diagram illustrating the effect of graded stretching (in the direction of the arrows) of a reticular matrix on the shape of mesenchymal cells.

Let us consider the two main intrinsic sources of morphogenetic tensions—expansion and contraction. On the left side (A_0 B_0) are sectors of a tissue possessing a continuous reticulated matrix and containing a circumscribed area (dark circle) in which some localized developmental process goes on, as follows. The lower half represents a case in which the marked area expands relative to its surroundings, either by rapid proliferation or by deposition of large amounts of ground substance or, in the case of hollow organs, by distension.

This local expansion creates in the surrounding continuum tensions that are oriented circumferentially, declining in strength with increasing distance from the border. The fibrous matrix assumes a corresponding pattern as depicted, and this in turn imparts itself to the cells. Cell movements and cell divisions will follow tangential courses. It is evident that the familiar concentric architecture of the mesenchymal sheaths, coats, tunics, and membranes around both solid organs and hollow ducts and vesicles finds a natural explanation on this basis. Essentially the same effect is observed if the surrounding tissue contracts against a rigid core, as in the formation of foreign-body capsules.

A wholly different pattern results if a tissue contains a focal area that contracts. In this case, illustrated in the top row of diagrams, the matrix adhering to the shrinking area is subject to radial tensions, and the surrounding tissue will thus be gathered into a starshaped pattern. Cell movements and divisions will be directed at the area, instead of around it as in the preceding example. Many embryonic invaginations (*e.g.* blastopore, neural tube, nasal pits) seem to take their origin from such localized contractile centers in the surface coat of the germ.

It was very illuminating to find that certain biochemical activities attending growth and differentiation may lead to the same type of expansion and contraction patterns. The expansion type, for instance, is observed around areas with proteolytic activity. When a local process dissolves part of an erstwhile coherent protein network, the remaining portion, owing to its elastic tension, retracts from the hole and assumes a pattern like that shown in Figure, much like a pierced cobweb. A contraction pattern such as in Figure, on the other hand, arises around actively proliferating centers through the following chain of events. Cells in proliferation, and apparently only then, liberate agents with strong dehydrating potency.

The result is a progressive condensation of the surrounding fibrous phase of the medium. Such local condensation in a rigid tissue frame generates tensions, which in turn orient the condensing system toward the centre of shrinkage. The familiar ray-shaped growth of ordinary tissue cultures in blood plasma is due to this effect. Analogous configurations can be recognized in many embryonic formations.

By its radial orientation such a contraction pattern can evidently guide peripheral cells toward the centre. This is one of the major mechanisms by which cells are drawn toward distant destinations, as if "attracted." Its morphogenetic importance becomes particularly impressive when several such centers act on a common matrix, where they create interconnections of great regularity among the otherwise independent centers. As is readily understood, two contractile centers produce the greatest amount of shrinkage, hence tension, and thereby orientation of the matrix along their connecting line; this automatically establishes a bridge for cell traffic between the centers. Three centers determine a triangle, and so forth.

In principle, these experiments have solved the problem of orientation in cell movement and growth. In acknowledging mechanical stress as a major morphogenetic factor, two reservations must be borne in mind. Firstly, tension is not the only agent capable of orienting tissue structure. Any other physical force that is capable of affecting the orientation and aggregation of polar molecules (electrostatic fields, electrophoresis, streaming, etc.) may have comparable effects. Secondly, stress is decidedly not a determining factor incytodifferentiation. Though it seems to be prerequisite for some cell types in order that they may be able to carry out terminal transformations for which they are already predisposed by previous differentiation, it is not the cause of such differentiation. Without stretch, myoblasts cannot transform into muscle fibres, but stretch itself neither makes a cell into a myoblast nor ever transforms a non-myoblast into a muscle fibre, as has once been claimed.

An interesting combination effect of both contractile tension and proteolysis has recently been observed in the healing of nerve stumps. Tension transmitted from stump to stump through an intervening blood clot, intensified by the syneretic contraction of the clot itself, orients the fibrin network in the latter in a prevailingly longitudinal direction. The oriented strands tend to fuse and thus become heavier than their irregularly arranged cross links. Then proteolytic enzymes appear on the scene and rout out the cross links, while leaving the straight linear bundles standing. Thereby preferentially linear order becomes strictly linear.

The main value of these varied results is that they demonstrate in principle how chemical activity (proteolysis, dehydration) can engender spatial order of a high degree, although it has itself no space order. Through its physical and chemical by-products, plain growth thus leads to ordered growth. As the order, arrangement, and consequent exposure of the growing tissue reciprocally affect the condition, manner, and hence, rate of its further growth, we realise how complex and diverse are the factors that contribute to the growth of a given tissue. Some of the factual sequences are summarized in Figure. Many more are still unexplored.

Carrying this exploration of the "microtechnology" of development, *i.e.* of "Entwicklungs-mechanik" in its original sense, for-ward on an intensified scale will go far in rectifying the current rather abstract and often patently unrealistic concepts of "differential growth."

But even the existing knowledge, of which we have presented some samples, is sufficiently compelling to discredit the growing fashion of reducing, on paper at least, problems of differential growth to terms of simple chemical reaction rates, with "growth stimulators" and "growth inhibitors" called upon to do the "regulating."

ELABORATION OF GROWTH PATTERNS

According to the preceding sections, differential growth is simply a corollary of differentiation, and differences in growth rates result from a great variety of causes. As such, we have singled out the unequal depletion of reproductive protoplasm in the building of specialized differentiation products in different cell types, the various inequalities arising from shifts, aggregations and dispersals of cell groups in accordance with properties acquired during differentiations, the resultant differences of geometric configuration, and hence, exposure of different parts of a tissue, with the ensuing confinement of proliferative activity to circumscribed zones of varying sizes, the different fates of differentiation products, which are retained in some tissues but extruded in others, and other related disparities contingent upon the divergent courses of differentiation.

Such inequalities in the growth pattern can, in turn, secondarily influence the setting for subsequent steps of differentiation, and the resulting changes will further modify the growth pattern. The actual elaboration of the mature body thus appears as a complex sequence of transforming patterns in which the contributions of differentiation and differential growth are intimately interlocked. Basically, however, differential growth is not an agent but merely a product and index of differentiation.

This realization is of the utmost importance when it comes to interpreting the manifold general non-localized influences which modify the growth process during later developmental stages, and which are impressively illustrated by the diversity of phenotypic variants that can be obtained from genetically identical germs: giants, dwarfs, duplications, defects, disproportions, excesses, and suppressions. Since these can often be directly related to aberrations of nutrition, hormones, oxygen supply, physical factors (radiations, pressure), and even climatic changes, all of these agents have at one time or other been claimed as the "dominating" factors in the growth process. We realise now that any factor that has an effect on the physico-chemical setting in which growth occurs may thereby alter the manifest form which the growing process will take. But it is evident that such actions will be in the nature of merely modifying already established patterns rather than instituting wholly new ones. Consequently an agent applied systemically (nontopically) may accentuate or reduce inherent differentials but cannot create differentials not yet in existence. A few illustrations may be in order.

As the developmental realization of the primordial growth patterns of the germ progresses, small initial differences among parts become greatly amplified. Minor deviations from normal become exaggerated. Therefore, if each part were to keep on growing independently of the other parts, the harmony of development would soon be endangered. The setting up of new systemic growth controls coordinating the separate parts thus becomes vital for harmonious development. This is achieved through the humoral, more specially the circulatory, systems for effective and rapid diffusion of substances.

But we understand now that a humoral pool could exercise no discriminative control over the various parts supplied by it unless these parts had already previously acquired different response qualities. For instance, the fraction of a common supply that a given part will be able to secure is a function of the accessibility of the part to the circulatory channels; this relation, however, is entirely a matter of previous differentiation and growth of both the part and the vessels. Similarly, the capacity of a part electively to remove, concentrate, utilize, or store certain components from a common pool only reflects affinities of its cells acquired during their prior biochemical specialization; affinity for iodine in the thyroid cell, for vitamin A in the sensory cell of the retina, etc.

Or if one cell group responds to a circulating hormone in a manner or degree different from that of another cell group, equally exposed, the two must have been already different constitutionally at the time the hormone appeared on the scene. The case is no different from that of selective drug response of mature cells, which is likewise based on the specific sensitivity of the reacting targets.

Since no nutrient or hormone or other chemical of ubiquitous distribution can make a cell render any contribution for which that cell has not been predisposed by its previous differentiation, none of those agents can be rated as primary causes of differential cell behaviour. But they do affect the subsequent elaboration and final proportions of the growth pattern in a variety of ways. A general deficiency of nutrients, oxygen, or specific growth accessories, for instance, hits different growing parts differentially, depending on their respective metabolic rates, margins of tolerance, reserves, substitutive capacities, and the like.

Faced with deficiencies, they compete for the available requisites in what Roux has described as the "struggle among parts." One part's loss may either be another part's gain, or may doom other parts, depending on the mode and degree of their mutual dependence. The production of differential growth defects by different vitamin deficiencies may be pointed out as a most instructive example.

But the lessons from such deficiencies are not reversible. While increasing the supply of growth essentials from deficient to adequate levels will benefit growth correspondingly, a further increase to levels of overabundance does not keep on enhancing growth indefinitely. A limit will be reached when the production plant works at full capacity, the maximum rate of production being determined by the intrinsic rates of the component processes, which, because of the assembly-line character of growth, are limited by the slowest links in the chains. A fuller discussion of these problems has been ably presented in other chapters of this book.

Of course, since what we call the "normal" level of nutrients, vitamins, hormones, and other growth requisites still falls short of the attainable optimum, the growth potentialities of the organism are "normally" not fully realized. Raising such growth factors to supranormal dosages can therefore occasionally have the spectacular effect of bringing out latent formations which under "normal" circumstances would have remained unrealized. For example, overdosing a female opossum with sex hormones causes the formation of prostate glands, a capacity inherent in the urogenital tract but failing to come to expression at "normal" hormone levels.

To the various humoral principles thus far recognized as secondary regulators of growth will perhaps have to be added an even more general category, namely, systems of the antigen-antibody type consisting of organ-specific discharges and their molecular counterparts. The case for this assumption has been presented on an earlier occasion, together with supporting experimental evidence. In the most condensed version, it is as follows. The specific key molecules which arise progressively in the course of differentiation act as templates ("positives") in the intracellular reproduction of more of their kind. At the same time they cause the formation of molecular species of complementary shape ("negatives"), which may or may not be an integral step in the reproductive process itself (double reversal). Complementary pairs could inactivate each other by conjugation. The amount of free active growth catalysts in a cell would thus vary with the excess of one type over the other.

Now, if we assume that the "positives" are retained in the cell (*e.g.* because of large dimensions), while the "negatives" can escape, circulate, and enter other cells (*e.g.* because of smaller size), we can see the general outlines of a highly specific and sensitive mechanism of growth regulation, which would coordinate all cells of common character into a single operational system, no matter how much they are scattered. Local reduction (or augmentation) of growth in any part of the system would imply a corresponding decline (or increase) of both "positives" and "negatives" in that locality, and because of their easier diffusibility, of the "negatives" only throughout the system. Since systemic reduction of the "negatives" means an increased ratio of active over conjugated "positives," a compensatory increase of growth throughout the rest of the system will ensue; and conversely, enhancement of growth in one part will automatically entail depression of growth in all other members of the system. Thus the total growth of a given system would be held under joint control of all its components.

That some such mechanisms exist is becoming increasingly evident from a study of compensatory growth reactions. That these mechanisms operate, at least in part, after the fashion of immune reactions is indicated by experiments in which organ-specific antibodies were shown to have selective effects on the growth of the homologous organs. But the precise mode of operation here sketched by way of illustration is sheer supposition.

Whatever its nature may eventually turn out to be, one must remember that it constitutes merely another contributing factor to the highly intricate system of dependencies and interactions which control the growth process. It is no more the master key to differential growth than are any of the other factors discussed before.

What our whole discussion adds up to is that there is no single master clue to the problems of differential growth or of growth in general. The measurable growth of different parts is so intimately dependent on their peculiar configurations and cellular differentiations, which themselves differ sharply and qualitatively, that a purely quantitative comparison of different growth processes is a deliberate abstraction. There are instances where such abstraction has led to the recognition of striking regularities capable of mathematical formulation (*e.g.* allotropic growth.

Whether these regularities are due to the fact that the compared systems resemble each other closely enough in their mode of growth to be essentially commensurable, or to the existence of some undefined principle of organismic nature, exercising overall control over the multiplicity of heterogeneous components involved in growth, can only be decided after a detailed analysis of the underlying phenomena. At any rate, a purely formal treatment of growth, as is often attempted through the interpretation of growth curves, is only a valuable guide to and supplement of, but never a substitute for, a precise analysis of the different forms in which growth manifests itself.

There can be no research on growth as such. We can only study growing objects. And different growing objects follow different methods. The growth of a differentiating metazoan introduces aspects not presented by the growth of a bacterial culture, and the growth of a higher plant has still other peculiarities. As we have pointed out repeatedly, each individual tissue and organ has its peculiar mode of growth. To know growth we must therefore first break down each one of its manifestations into its constituent elementary processes and then study these and describe them in objective terms. This is a long way to go, but there is no short cut.

In our present primitive stage of knowledge impatient attempts to formulate a general and universal theory of growth seem to have little chance of success. Those general theories that are sporadically being advanced are usually "general" only in the sense that they are first derived from some rather special segment of the growth problem and then broadly generalized by proclamation or mere implication. Consequently they may be sound and pertinent within the area of their derivation but gratuitous and invalid in their illegitimate extensions. Oversimplification of facts and overgeneralization in theory are the main causes of the existing discordance among the several contending concepts of growth. Unless we desist from those practices the confusion is bound to grow, in spite of all the excellent experimental work being done in various areas.

Growth is not a simple and unitary phenomenon. Growth is a word, a term, a notion, covering a variety of diverse and complex phenomena. It is not even a scientific term with defined and constant meaning, but a popular label that varies with the accidental traditions, predilections, and purposes of the individual or school using it. It has come to connote all and any of these: reproduction, increase in dimensions, linear increase, gain in weight, gain in organic mass, cell multiplication, mitosis, cell migration, protein synthesis, and perhaps more.

It is gravely inconsistent to apply the most exacting standards of precision to our research data and then proceed to mix into their description and interpretation such vague terminology as this. The mixture can be no more precise than its vaguest ingredient. To deal with growth as an entity, which can be "activated," "stimulated," "retarded," or "suppressed," is only part science, and for the other part, fiction. The less we let our work and thoughts be misled by the delusion that "growth" is basically but a simple elementary process, like a "bimolecular reaction," the faster will be our progress toward true insight into the real mechanisms of development.

BASIC STEPS OF PLANT TISSUE CULTURE

Some basic steps are followed in plant tissue culture which is described below.

Explant Selection

Explant is the plant fraction used for tissue culture. Explant from healthy and young part of the plant is used. Parenchymas from stems, rhizomes, tubers, roots are easily accessible and respond quickly to culture condition.

Sterilisation Elimination

Sterilisation means elimination of all living organisms (microbes). The tissue culture is carried out in completely aseptic condition for which there is a need to properly sterilise all the glasswares culture media and explants. The explants are surface sterilised by repeated washing in sterile water and by using disinfectants such as mercuric chloride, hydrogen peroxide, etc. The glassware, culture media and other instruments are sterilised in an instrument called autoclave where sterilisation is done in steam under high pressure or in a pressure cooker.

Preparation of Nutrient or Culture Medium

The culture medium is prepared in aseptic condition. The basic constituents of any culture medium are:

- *Inorganic nutrients:* Inorganic nutrients include macronutrients as salts of nitrogen, phosphorous, potassium, calcium, magnesium and sulphur and micronutrients like boron, molybdenum, copper, zinc. Iron and chloride

Source of Carbon

Sucrose is used as the source of carbon. Several growth hormones like 2,4,1), cytokinins-benzylaminopurine, kinetin, myoinositol, 1AA, XAA arc used, vitamins like nicosstinic acid, pyridoxinc-IICL are added to the medium. After the addition of all the ingredients in appropriate proportion, agar-agar is added to prepare a solid medium. In some types of culture (like root culture) liquid medium is used (no agar-agar used).

Process of Inoculation

Inoculation is the process of the transfer of explant to suitable nutrient medium contained in culture vessels. This is done in sterile condition either in an inoculation chamber or laminar flow. After the inoculation the culture vessels are maintained in controlled temperature and light. The suitable temperature, for tissue culture ranges between 18-25°C.

Callus Configuration and its Culture

A callus is an amorphous mass of loosely arranged thin walled parenchyma cells developing from proliferating cells of parent tissue (Dodos and Roberts, 1985). The nutrient medium supplemented with auxins induces cell division and soon the upper surface of explant is covered by callus. The callus has the biological potential to develop normal root, shoots and ultimately forming a plant. Callus is formed through three developmental stages: induction, cell division and differentiation. Callus formation is governed by the source of explant, nutritional composition of the medium and environmental factors.

During the induction period the metabolic rate of cells is stimulated. Owing to increased metabolic rate cells enter to cell division stage. In the third phase, cellular differentiation and expression of certain metabolic pathways start leading to secondary products. When callus are grown on a nutrient medium for a long time, it becomes essential to subculture it in fresh media.

Development of Organogenesis

Organogenesis means the development of organs like root, shoot and leaves (but not embryo). Organogenesis starts with stimulation caused by the chemical of medium, substances carried over from the original explants and endogenous compounds produced by the culture. Organogenesis can be induced with the application of varying proportions of auxin and cytokinin. Skoog and Miller (1957) demonstrated that a high ratio of auxin: cytokinin (3: 0.02) stimulated root formation in tobacco callus but a low ratio of the same (3: 0,2) induced shoot formation.

Formation of Somatic Embryogenesis

This is the process of inducing embryo formation from somatic cells of cultured plant tissue. The embryo thus developed is knowTi as embryoids. Two different nutritional media are required to obtain embryoids. First medium contains auxin to initiate embryogenic cells. Second medium lacks auxin or has reduced level of auxins for subsequent development of embryonic cells into embryoids and plantlets.

The embryogenic cells pass through three different stages, *e.g.*, globular, heart shaped and torpedo shaped to form embryoids. Some plants in which somatic embryogenesis has been induced *in vitro* are Atropa belladona, Brassica oleracea, Carica papaya, Coffea arabica, Citrus cinensis, Daucus carroia, Nicotiana tobacum, etc.

Somaclonal Variations: In 1981, P.J Larkins and W. R. Scowcrott at the Division of Plant Industry, C.S.I.R.O. Australia gave the name somaclonal variation to genetic variability generated during tissue culture.

Though the cultured tissues are grown from single explant, over a long period of maintenance genetic variabilities are marked in the cultures.

This may be due to:

- Reflection of heterogeneity between cells and explant tissues,
- A simple representation of spontaneous mutation rate or
- Activation by culture environment of transposition of genetic materials.

Somaclonal variants of leaf callus culture of Solanum tuberosum have shown characters like disease resistance, variations in maturity dates of tubers, yield and shape.

Cell suspension cultures: Cell suspension is prepared by transferring a fragment of callus (about 500 mg) to liquid medium (500 ml) and agitating them aseptically in a shaker to make the cells free. The suspension then includes single cell, cell aggregates, and residual including dead cells. A good suspension contains high proportion of single cells.

Cell suspension cultures have many advantages over the callus cultures as below:

- The cell suspension can be pipetted.
- They are less heterogeneous.
- They can be cultured in volumes up to 1,500 litres.
- They can be subjected to more stringent environmental controls.
- They can be manipulated for production of natural products.

HISTORY OF TISSUE CULTURE TECHNIQUES

The use of tissue culture techniques initiated by Gottlieb Haberlandt in 1902 in an attempt mengkulturkan hair cells of the leaf mesophyll tissue monocot plant. But the effort failed because the cells do not have cleavage, it was alleged failures because they do not use growth regulator substances needed for cell division, proliferation and induction of the embryo. In the year 1904, Hannig planting embryos isolated from several plant crucifers. In 1922, Knudson and separately each Robbin conduct investment business and culture of orchid seedlings root tip. After the 1920s, the discovery and development of tissue culture techniques continues. The following table shows the historical development of the field of plant tissue culture which was adapted from various sources. Important discoveries in the history of plant tissue culture.

Year important findings:

1902: *Haberlandt*: The first person who tried to isolate and mengkulturkan monocot plant tissue, but failed

1922: *Knudson*: mengecambahkan orchid seeds

Blumenthal and Meyer 1924: The formation of callus from carrot roots eksplan

1929: *Laibach and Hered*: Culture of embryos to overcome the incompatibility in plants Linum spp.

1934: Culture in vitro oflGautheret woody plant tissue kambium and shrubs, but failed.

White: Successfull culture of tomato roots in a long time

Kogl et.al: Identification of thel first plant hormone, IAA, for the elongation of cells.

1936: *LaRue*: Culture of embryos in some plants Gymnospermae

1939: *Gautheret*: Successfully growing culture kambium carrot and tobacco plants

1941: *Overbeek*: The use of coconut water for the young embryo culture menumbuhkam Datura plants

1944: The first in vitro culture of tobacco plants to study the formation of buds adventif

Skoog and Tsui 1948: The formation of shoots and roots of tobacco adventif

1949: *Nitsch*: in vitro culture of fruit plants

1952: *Morel and Martin*: meristem culture to obtain a free Dahlia plant viruses. The firstl success of micro-called grafting.

1953: *Tulecke*: Kalus haploid pollen plants of Ginkgo biloba *Miller* 1955: The discovery of the structure and synthesis of kinetin

Skoog and Miller 1957: Menemuan that the formation of roots and shoots in comparison tergatung culture auksin: cytokines

1958: *Maheswaril and Rangaswamy*: Regeneration of somatic embryos Citrus nuselus ovul

Reinert and Steward: Growth and development of carrot suspension cultures

λ*Cocking* 1960: enzymatic degradation of cell walls to get protoplas

Morel: λ Vegetative Propagation of orchids by meristem culture

1962: *Murashige and Skoog*: Development of MS medium

1964: *Guha and Maheswari*: The discovery of the first haploid plants by androgenesis Datura plant

Erickson and Jonassen 1969: Isolation protoplas of cell suspension Hapopappus

1970: *Power*: Figaro protoplas

1977: *The success of T-DNA integration in plantslChilton*

Noguchi et al: Planting of tobacco cells in Bioreaktor capacity of 20 000 L.

1978: *Melchers et al*: Somatic hybridization between tomato and potatol plants

Tabata et al: Production of shikonin on the industrial scale celll culture

Zimmermann 1982: Figaro protoplas electrically (Electrofusion)

Mitsui Petrochemicals 1983: The first secondary metabolite production in industrial scale through suspension culture at the plant Lithospermum spp.

1985-1990: Development of gene transfer in rapidly growing crops, such as the use of Agrobacterium, particle bombardment (gene gun), electroporasi, mikroinjeksi.

Development of genetic and metabolicλ1990 - engineering challenges. growing rapidly Marketing products of geneticλ engineering. The in vitro techniques were developed initially to demonstrate the totipotency of plant cells predicted by Haberlandt in 1902. Totipotency is the ability of a plant cell to perform all the functions of development, which are characteristic of zygote, *i.e.*, ability to develop into a complete plant. In 1902, Haberlandt reported culture of isolated single palisade cells from leaves in Knop's salt solution enriched with sucrose. The cells remained alive for up to 1 month, increased in size, accumulated starch but failed to divide. Efforts to demonstrate totipotency led to the development of techniques for cultivation of plant cells under defined conditions.

Haploid plants from pollen grains were first produced by Maheshwari and Guha in 1964 by culturing anthers of Datura. This marked the beginning of anther culture or pollen culture for the production of haploid plants. The technique was further developed by many workers, more notably by JP. Nitch, C. Nitch and coworkers. These workers showed that isolated microspores of tobacco produce complete plants. Plant protoplasts are naked cells from which cell wall has been removed.

In 1960, Cocking produced large quantities of protoplasts by using cell wall degrading enzymes. The techniques of protoplast production have now been considerably refined. It is now possible to regenerate whole plants from protoplasts and also to fuse protoplasts of different plant species. In 1972, Carlson and coworkers produced the first somatic hybrid plant by fusing the protoplasts of Nicotiana glauca and N. langsdorfii. Since then many divergent somatic hybrids have been produced. A successful establishment of callus cultures depended on the discovery during mid-thirties of IAA (idole-3-acetic acid), the endogenous auxin, and of the role of B vitamins in plant growth and in root cultures.

The first continuously growing callus cultures were established from cambium tissue in 1939 independently by Gautheret, White and Nobecourt. The subsequent discovery of kinetin by Miller and coworkers in 1955 enabled the initiation of callus cultures from differentiated tissues. Shoot bud differentiation from tobacco pith tissues cultured in vitro was reported by Skoog in 1944, and in 1957 Skoog and Miller proposed that root-shoot differentiation in this system was regulated by auxin-cytokinin ratio. The first plant from a mature plant cell was regenerated by Braun in 1959. Development of somatic embryos was first reported in 1958- 1959 from carrot tissues independently by Reinert and Steward. Thus within a brief period, the tissue culture techniques have made a great progress. From the sole objective of demonstrating the totipotency of differentiated plant -cells, the technique now finds application in both basic and applied researches in a number of-fields of enquiry.

ENVIRONMENTAL TECHNIQUES OF PLANT TISSUE

The term "Environment" is defined as our surroundings which includes the abiotic component (the non living) and biotic component (the living) around us. The abiotic environment includes water, air and soil while the biotic environment consists of all living organisms – plants, animals and microorganisms. Environmental pollution broadly refers to the presence of undesirable substances in the environment which are harmful to man and other organisms. In the past decade or two, there has been a significant increase in the levels of environmental pollution mostly due to direct or indirect human activities. The major sources of environmental pollution are –Industries, Agricultural sources (mainly rural area), anthropogenic sources (man related activities mainly in urban areas), biogenic sources etc. The pollutants are chemical, biological and physical in nature. The Chemical pollutants include-gaseous pollutants (hazardous gases like sulfur dioxide, nitrogen oxide), toxic metals, pesticides, herbicides toxins and carcinogens

Etc. The physical pollutants are- heat, sound, radiation, and radioactive substances. The pathogenic organisms and some poisonous and dangerous biological products are the biological pollutants. Controlling the environmental pollution and the conservation of environment and biodiversity and controlling environmental pollution are the major focus areas of all the countries around the world.

In this context the importance and impact of biotechnological approaches and the implications of biotechnology has to be thoroughly evaluated. There have been serious concerns regarding the use of biotechnology products and the impact assessment of these products due to their interaction with the environmental factors. A lobby of the environmentalists have expressed alarm on the release of genetically engineered organisms in the atmosphere and have stressed on thorough investigation and proper risk assessment of theses organisms before releasing them in to the environment. The effect of the effluents from biotechnological companies is also a cause of concern for everyone. The need of the hour is to have a proper debate on the safety of the use of the biotechnological products.

The efforts are not only on to use biotechnology to protect the environment from pollutionbut also to use it to conserve the natural resources. As we all know that microorganisms are known natural scavengers so the microbial preparations (both natural as well as genetically engineered) can be used to clean up the environmental hazards.

DEVELOPMENT OF ALTERNATE CLEANER TECHNOLOGIES USING BIOTECHNOLOGY

Biotechnology is being used to provide alternative cleaner technologies which will help to further reduce the hazardous environmental implications of the traditional technologies. *E.g.* some Fermentation technologies have some serious environmental implications. Various biotechnological processes have been devised in which all nutrients introduced for fermentation are retained in the final product, which ensures high conversion efficiency and low environmental impact.

In paper industry, the pulp bleaching technologies are being replaced by more environmentally friendly technologies involving biotechnology. The pulp processing helps to remove the lignin without damaging valuable cellulosic fibres but the available techniques suffer from the disadvantages of high costs, high energy use and corrosion. A lignin degrading and modifying enzyme (LDM) was isolated from Phanerochaete chrysosporum and was used, which on one hand, helped to reduce the energy costs and corrosion and on the other hand increased the life of the system. This approach helped in reducing the environmental hazards associated with bleach plant effluents. In Plastic industry, the conventional technologies use oil based raw materials to extract ethylene and propylene which are converted to alkene oxides and then polymerized to form plastics such as polypropylene and polyethylene. There is always the risk of these raw materials escaping into the atmosphere thereby causing pollution. Using biotechnology, more safer raw materials like sugars (glucose) are being used which are enzymatically or through the direct use of microbes converted into alkene oxides.*e.g.* Methylococcus capsulatus has been used for converting alkene into alkene oxides.

BIOREMEDIATION

Bioremediation is defined as 'the process of using microorganisms to remove the environmental pollutants where microbes serve as scavengers. The removal of organic wastes by microbes leads to environmental cleanup. The other names/terms used for bioremediation are biotreatment, bioreclamation, and biorestoration. The term "Xenobiotics" (xenos means foreign) refers to the unnatural, foreign and synthetic chemicals such as pesticides, herbicides, refrigerants, solvents and other organic compounds. The microbial degradation of xenobiotics also helps in reducing the environmental pollution.

Pseudomonas which is a soil microorganism effectively degrades xenobiotics. Different strains of Pseudomonas that are capable of detoxifying more than 100 organic compounds (*e.g.* phenols, biphenyls, organophosphates, naphthalene etc.) have been identified. Some other microbial strains are also known to have the capacity to degrade xenobiotics such as Mycobacterium, Alcaligenes, Norcardia etc.

FACTORS AFFECTING BIODEGRADATION

The factors that affect the biodegradation are: the chemical nature of xenobiotics, the concentration and supply of nutrients and O_2, temperature, pH, redox potential and the capability of the individual microorganism. The chemical nature of xenobiotics is very important because it was found out that the presence of halogens *e.g.* in aromatic compounds inhibits biodegradation. The water soluble compounds are more easily degradable whereas the presence of cyclic ring structure and the length chains or branches decrease the efficiency of biodegradation. The aliphatic compounds are more easily degraded than the aromatic ones.

BIOSTIMULATION

It is a process by which the microbial activity can be enhanced by increased supply of nutrients or by addition of certain stimulating agents like electron acceptors, surfactants etc.

BIOAUGMENTATION

It is possible to increase biodegradation through manipulation of genes *i.e.* using genetically engineered microorganisms and by using a range of microorganisms in biodegradation reaction.

Depending on the method followed to clean up the environment, the bioremediation is carried out in two ways:

In Situ Bioremediation

In situ bioremediation involves a direct approach for the microbial degradation of xenobiotics at the site of pollution which could be soil, water etc. The adequate amount of essential nutrients is supplied at the site which promotes the microbial growth at the site itself. The in situ bioremediation is generally used for clean up of oil spillages, beaches etc.

There are two types of in situ bioremediation:

1. *Intrinsic bioremediation*: The microorganisms which are used for biodegradation are tested for the natural capability to bring about biodegradation. So the inherent metabolic ability of the microorganisms to degrade certain pollutants is the intrinsic bioremediation. The ability of surface bacteria to degrade a given mixture of pollutants in ground water is dependent on the type and concentration of compounds, electron acceptor and the duration of bacteria exposed to contamination. Therefore, the ability of indigenous bacteria degrading contaminants can be determined I laboratory by using the techniques of plate count and microcosm studies. The conditions of site that favour intrinsic bioremediation are ground water flow throughout the year carbonate minerals to buffer acidity produced during biodegradation, supply of electron acceptors and nutrients for microbial growth and absence of toxic compounds.
2. *Engineered in situ bioremediation*: When the bioremediation process is engineered to increase the metabolic degradation efficiency (of pollutants) it is called engineered in situ bioremediation. This is done by supplying sufficient amount of nutrients and oxygen supply, adding electron acceptors and maintaining optimal temperature and pH. This is done to overcome the slow and limited bioremediation capability of microorganisms.

Advantages of in Situ Bioremediation

- The method ensures minimal exposure to public or site personnels.
- There is limited or minimal disruption to the site of bioremediation.
- Due to these factors it is cost effective.
- The simultaneous treatment of contaminated soil and water is possible.

Disadvantages of in Situ Bioremediation

- The sites are directly exposed to environmental factors like temperature, oxygen supply etc.
- The seasonal variation of microbial activity exists.
- Problematic application of treatment additives like nutrients, surfactants, oxygen etc.
- It is a very tedious and time consuming process.

Ex-situ Bioremediation

In this the waste and the toxic material is collected from the polluted sites and the selected range of microorganisms carry out the bioremediation at designed place. This process is an improved method over the in situ bioremediation method.

On the basis of phases of contaminated materials under treatment ex-situ bioremediation is classified into two:

1. *Solid phase treatment*: This system includes land treatment and soil piles comprising of organic wastes like leaves, animal manures, agricultural wastes, domestic and industrial wastes, sewage sludge, and municipal solid wastes. The traditional clean-up practice involves the informal processing of the organic materials and production of composts which may be used as soil amendment. Composting is a self heating, substrate-dense, managed microbial system which is used to treat large amount of contaminated solid material. Composting can be done in open system *i.e.* land treatment and/or in closed treatment system. The hazardous compounds reported to disappear through composting includes aliphatic and aromatic hydrocarbons and certain halogenated compounds. The possible routes leading to the disappearance of hazardous compounds include volatilization, assimilation, adsorption, polymerization and leaching.
2. *Slurry phase treatment*: This is a triphasic treatment system involving three major components- water, suspended particulate matter and air. Here water serves as suspending medium where nutrients, trace elements, pH adjustment chemicals and desorbed contaminants are dissolved. Suspended particulate matter includes a biologically inert substratum consisting of contaminants and biomass attached to soil

matrix or free in suspending medium. The contaminated solid materials, microorganisms and water formulated into slurry are brought within a bioreactor *i.e.* fermenter. Biologically there are three types of slurry-phase bioreactors: aerated lagoons, low shear airlift reactor, and fluidized-bed soil reactor. The first two types are in use of full scale bioremediation, while the third one is in developmental stage.

Advantages of Ex-situ Bioremediation

- As the time required is short, it is a more efficient process.
- It can be controlled in a much better way.
- The process can be improved by enrichment with desired and more efficient microorganisms.

Disadvantages of Ex-situ Bioremediation

- The sites of pollution remain highly disturbed.
- Once the process is complete, the degraded waste disposal becomes a major problem.
- It is a costly process.

Several types of reactions occur during the bioremediation/microbial degradation:

- *Aerobic bioremediation*: When the biodegradation requires oxygen O_2 for the oxidation of organic compounds, it is called aerobic bioremediation. Enzymes like monooxygenases and dioxygenases are involved and act on aliphatic and aromatic compounds.
- *Anaerobic bioremediation*:This does not require oxygen O_2. the degradation process is slow but more cost effective since continuous supply of oxygen is not required.
- *Sequential bioremediation*: Some of the xenobiotic degradation requires both aerobic as well as anaerobic processes which very effectively reduces the toxicity *e.g.* tetrachloromethane and tetrachloroethane undergo sequential degradation.

Use of genetic engineering and genetic manipulations for more efficient bioremediation. In recent years, efforts have been made to create genetically engineered microorganisms (GEMs) to enhance bioremediation. This is done to overcome some of the limitations and problems in bioremediation.

These problems are:

- Sometimes the growth of microorganisms gets inhibited or reduced by the xenobiotics.
- No single naturally occurring microorganisms has the capability of degrading all the xenobiotics present in the environmental pollution.
- The microbial degradation is a very slow process.
- Sometimes certain xenobiotics get adsorbed on to the particulate matter of soil and thus become unavailable for microbial degradation.

As the majority of genes responsible for the synthesis of enzymes with biodegradation capability are located on the plasmids, the genetic manipulations of plasmids can lead to the creation of new strains of bacteria with different degradative pathways. In 1970s, Chakrabarty and his team of co-workers reported the development of a new strain of bacterium Pseudomonas by manipulations of plasmid transfer which they named as "superbug". This superbug had the capability of degrading a number of hydrocarbons of petroleum simultaneouslysuch as camphor, octane, xylene, naphthalene etc.

In 1980, United States granted the patent to this superbug making it the first genetically engineered microorganism to be patented. In certain cases, the process of plasmid transfer was used. *E.g.* The bacterium containing CAM (camphor degrading) plasmid was conjugated with another bacterium with OCT (octane degrading) plasmid. Due to non-compatibility, these plasmids cannot coexist in the same bacterium.

However, due to the presence of homologous regions of DNA, recombination occurs between these two plasmids which results in a single CAM-OCT plasmid giving the bacterium the capacity to degrade both camphor as well as octane. A new strain of Pseudomonas sp. (strain ATCC 1915) has been developed for the degradation of vanillate (which is a waste product from paper industry) and sodium dodecyl sulfate (SDS, a compound used in detergents).

BIOTECHNOLOGICAL METHOD TO REDUCE ATMOSPHERIC CARBON DIOXIDE (CO_2)

Carbon dioxide is the gas that is the main cause of green house effect and rise in the atmospheric temperature. During the past 100-150 years, the level of CO_2 has increased about 25% with an increase in the atmospheric temperature by about 0.5% which is a clear indication that CO_2 is closely linked with global warming. There is a steady increase in the CO_2 content due to continuous addition of CO_2 from various sources particularly from industrial processes. It is very clear that the reduction in atmospheric CO_2 concentration assumes significance.

Biotechnological methods have been used to reduce the atmospheric CO_2 content at two levels:

1. *Photosynthesis*: Plants utilize CO_2 during the photosynthesis which reduces the CO_2 content in the atmosphere. The equation for photosynthesis is:
   sunlight
   $$6CO_2 + 6H_2O \rightarrow C_6H_{12}O_6 + 6O_2$$
   Chlorophyll

   The fast growing plants utilize the CO2 more efficiently for photosynthesis. The techniques of micropropagation and synthetic

seeds should be used to increase the propagation of such fast growing plants.

Further, the CO_2 utilization can be increased by enhancing the rate of photosynthesis. The enzyme ribulose biphosphate carboxylase (RUBP-case) is closely linked with CO_2 fixation. The attempts are being made to genetically manipulate this enzyme so that the photosynthetic efficiency is increased.

Some microalgae like Chlorella pyrenodiosa, Spirulina maxima are known to be more efficient than higher plants in utilizing atmospheric CO_2 for photosynthesis and generate more O_2 than the amount of CO_2 consumed.

The growing of these microalgae near the industries and power plants (where the CO_2 emission in to atmosphere is very high) will help in the reduction of polluting effects of CO_2. Using genetic engineering, attempts are going on to develop new strains of these microalgae that can tolerate high concentrations of CO_2. A limited success has already been reported in the mutants ofAnacystis nidulans and Oocystis sp.

2. Biological Calcification- Certain deep sea organisms like corals, green and red algae store CO_2 through a process of biological calcification. As the $CaCO_3$ gets precipitated, more and more atmospheric CO_2 can be utilized for its formation. The process of calcification is as follows:

$$H_2O + CO_2 \rightarrow H_2CO_3$$
$$H_2CO_3 + Ca^{2+} \rightarrow CaCO_3 + CO_2 + H_2O$$

TREATMENT OF SEWAGE USING MICROORGANISMS

The sewage is defined as the waste water resulting from the various human activities, agriculture and industries and mainly contains organic and inorganic compounds, toxic substances, heavy metals and pathogenic organisms. The sewage is treated to get rid of these undesirable substances by subjecting the organic matter to biodegradation by microorganisms. The biodegradation involves the degradation of organic matter to smaller molecules (CO_2, NH_3, PO_4 etc.) and requires constant supply of oxygen. The process of supplying oxygen is expensive, tedious, and requires a lot of expertise and manpower. These problems are overcome by growing microalgae in the ponds and tanks where sewage treatment is carried out. The algae release the O_2 while carrying out the photosynthesis which ensures a continuous supply of oxygen for biodegradation.

TOOLS FOR STUDYING CELLS

Cell biologists would love to do what you just did—shrink down and actually see, touch, and hear the inner workings of cells. Because that's impossible, they've developed an ever–growing collection of approaches to study cellular innards from the outside. Among them are biochemistry, physical analysis, microscopy, computer analysis, and molecular genetics.

Using these techniques, researchers can exhaustively inventory the individual molecular bits and pieces that make up cells, eavesdrop on cellular communication, and spy on cells as they adapt to changing environments. Together, the approaches provide vivid details about how cells work together in the body's organs and tissues. We'll start by discussing the traditional tools of the trade—microscopes—then touch on the new frontiers of quantum dots and computational biology.

LIGHT MICROSCOPES: THE FIRST WINDOWS INTO CELLS

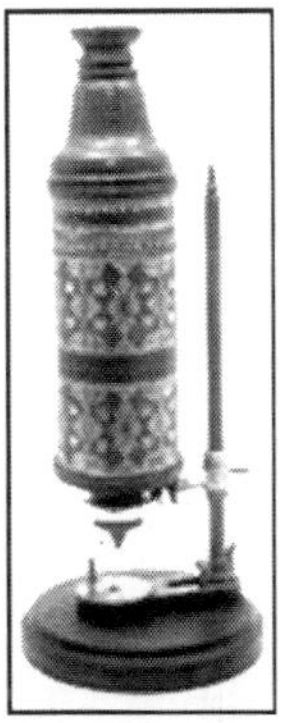

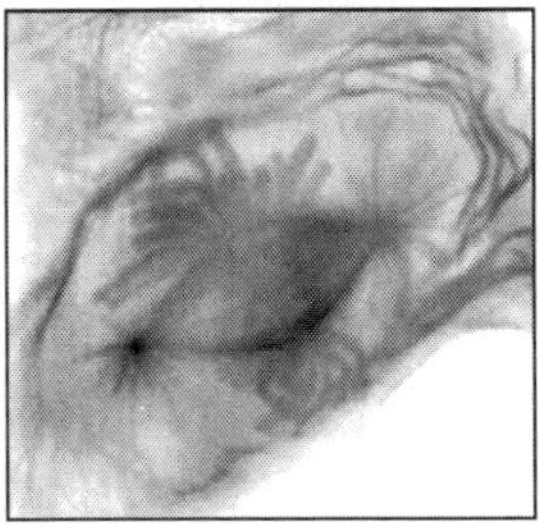

Scientists first saw cells by using traditional light microscopes. In fact, it was, looking through a microscope at a thin slice of cork, who coined the word "cell." He chose the word to describe the boxlike holes in the plant cells because they reminded him of the cells of a monastery.Scientists gradually got better at grinding glass into lenses and at whipping up chemicals to selectively stain cellular parts so they could see them better. By the late 1800s, biologists already had identified some of the largest organelles (the nucleus, mitochondria, and Golgi).Researchers using high–tech light microscopes and glowing molecular labels can now watch biological processes in real time. The scientists start by chemically attaching a fluorescent dye or protein to a molecule that interests them. The coloured glow then allows the scientists to locate the molecules in living cells and to track processes—such as cell movement, division, or infection—that involve the molecules.Fluorescent labels come in many colours, including brilliant red, magenta, yellow, green, and blue. By using a collection of them at the same time, researchers can label multiple structures inside a cell and can track several processes at once. The technicolour result provides great insight into living cells—and is stunning cellular art.

ELECTRON MICROSCOPES: THE MOST POWERFUL OF ALL

In the 1930s, scientists developed a new type of microscope, an electron microscope that allowed them to see beyond what some ever dreamed possible. The revolutionary concept behind the machine grew out of physicists' insights into the nature of electrons.

As its name implies, the electron microscope depends not on light, but on electrons. The microscopes accelerate electrons in a vacuum, shoot them out of an electron gun, and focus them with doughnut–shaped magnets. As the electrons bombard the sample, they are absorbed or scattered by different cell parts, forming an image on a detection plate.

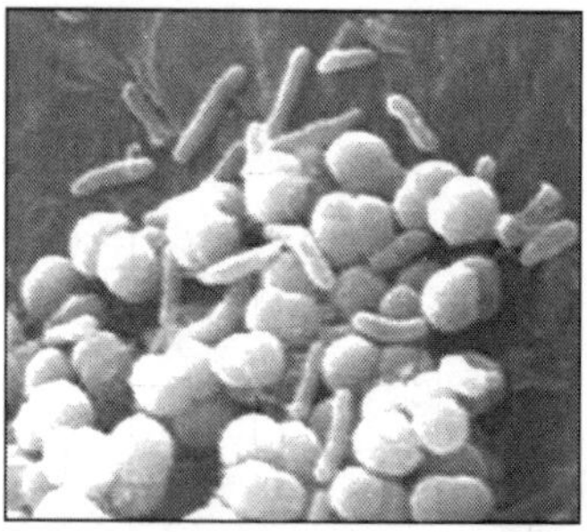

Although electron microscopes enable scientists to see things hundreds of times smaller than anything visible through light microscopes, they have a serious drawback: They can't be used to study living cells. Biological tissues don't survive the technique's harsh chemicals, deadly vacuum, and powerful blast of electrons.Electron microscopes come in two main flavours: transmission and scanning. Some transmission electron microscopes can magnify objects up to 1 million times, enabling scientists to see viruses and even some large molecules. To obtain this level of detail, however, the samples usually must be sliced so thin that they yield only flat, two–dimensional images. Photos from transmission electron microscopes are typically viewed in black and white.Scanning electron microscopes cannot magnify samples as powerfully as transmission scopes, but they allow scientists to study the often intricate surface features of larger samples. This provides a window to see up close the three–dimensional terrain of intact cells, material surfaces, microscopic organisms, and insects. Scientists sometimes use computer drawing programmes to highlight parts of these images with colour.

STUDYING SINGLE MOLECULES: CONNECTING THE QUANTUM DOTS

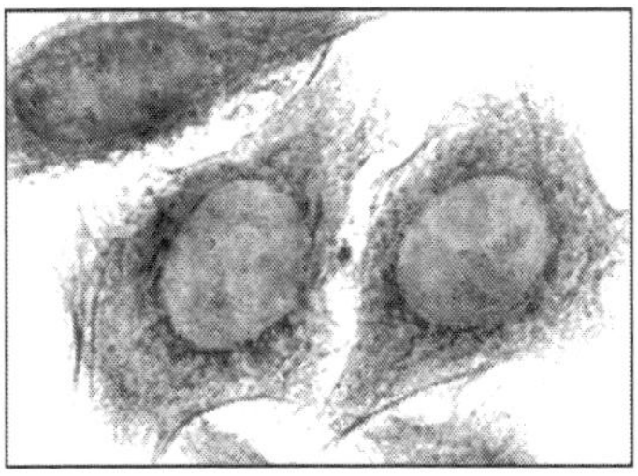

Dyes called quantum dots can simultaneously reveal the fine details of many cell structures. Someday, the technique may be used for speedy disease diagnosis, DNA testing, or analysis of biological samples. Whether they use microscopes, genetic methods, or any other technique to observe specific molecules, scientists typically flag every molecule of a certain type, then study these molecules as a group. It's rather like trying to understand a profession—say, teaching, architecture, or medicine—by tagging and observing all the workers in that profession simultaneously. Although these global approaches have taught us a lot, many scientists long to examine individual molecules in real time—the equivalent of following individual teachers as they go about their daily routines.

Now, new techniques are beginning to allow scientists to do just that. One technology, called quantum dots, uses microscopic semiconductor crystals to label specific proteins and genes. The crystals, each far less than a millionth of an inch in diameter, radiate brilliant colours when exposed to ultraviolet light.

Dots of slightly different sizes glow in different fluorescent colours—larger dots shine red, while smaller dots shine blue, with a rainbow of colours in between. Researchers can create up to 40,000 different labels by mixing quantum dots of different colours and intensities as an artist would mix paint. In addition to coming in a vast array of colours, the dots also are brighter and more versatile than more traditional fluorescent dyes: They can be used to visualize individual molecules or, like the older labeling techniques, to visualize every molecule of a given type.

Quantum dots promise to advance not only cell biology but also a host of other areas. Someday, the technology may allow doctors to rapidly analyse thousands of genes and proteins from cancer patients and tailor treatments to each person's molecular profile. These bright dots also could help improve the speed, accuracy, and affordability of diagnostic tests for everything from HIV infection to allergies. And, when hitched to medicines, quantum dots might deliver a specific dose of a drug directly to a certain type of cell.

COMPUTERS CLARIFY COMPLEXITY

Say you're hungry and stranded in a blizzard: If you eat before you seek shelter, you might freeze to death, but if you don't eat first, you might not have the strength to get yourself out of the storm. That's analogous to the decisions cells have to make every day to survive.

For years, scientists have examined cell behaviours—like the response to cold or hunger—one at a time. And even that they did bit by bit, labouriously hammering out the specific roles of certain molecules. This approach made it difficult or impossible to study the relative contributions of—and the interplay between—genes that share responsibility for cell behaviours, such as the 100 or so genes involved in the control of blood pressure. Now, computers are allowing scientists to examine many factors involved in cellular behaviours and decisions all at the same time.

The field of computational biology blossomed with the advent of high-end computers. For example, sequencing the 3.2 billion base pairs of the human genome, which was completed in 2003, depended on computers advanced enough to tackle the challenge.Now, state-of-the-art equipment and a wealth of biological data from genome projects and other technologies are opening up many new research opportunities in computer analysis and modeling. So, much as microscopes and biochemical techniques revolutionized cell biology centuries ago, computers promise to advance the field just as dramatically in this new century.

SCIENCE SCHISMS

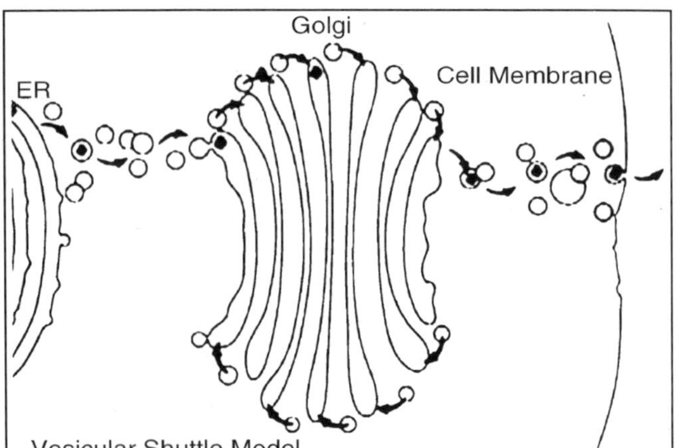

In living cells, material moves both forward (above arrows in figure) and backward (below arrows in figure) from each tip of the Golgi. For simplicity, we have illustrated forward movement only on the top and backward movement only on the bottom of the Golgi cartoon.One great thing about science is that you're allowed to argue about your work.To gain new information, scientists ask a lot of questions. Often, the answers spur more questions. The never–ending circle not only feeds curiosity; it also can lead to important and sometimes unexpected breakthroughs. Sometimes, scientists studying the same topic but using different experimental approaches come up with different conclusions.

Take the Golgi, for example. Think it's noncontroversial? The details of how this organelle forms inside your cells have kept two camps of researchers in a lively battle.

On one side of the debate is Graham Warren of Yale University School of Medicine in New Haven, Connecticut, who argues that the Golgi is an architectural structure that cannot be made from scratch. He believes that newly made proteins are packaged in the rough ER and are sent for further processing to a pre–existing structure (the Golgi) that is made up of different compartments. This is called the vesicular shuttle model.

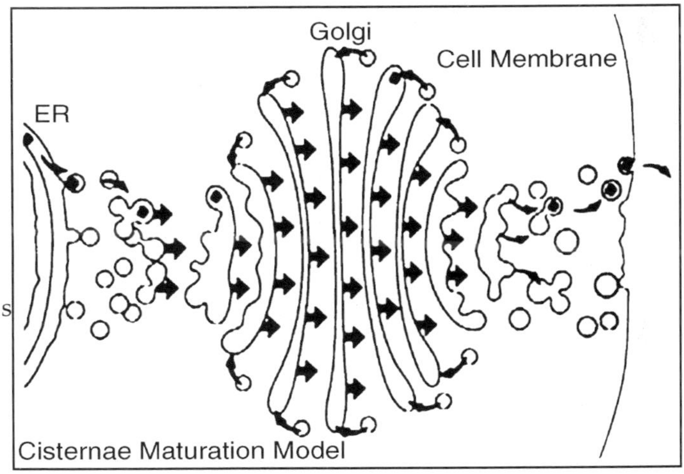

On the other side is Jennifer Lippincott–Schwartz of the National Institute of Child Health and Human Development (part of the National Institutes of Health) in Bethesda, Maryland. She says that the Golgi makes itself from scratch. According to her theory, packages of processing enzymes and newly made proteins that originate in the ER fuse together to form the Golgi. As the proteins are processed and mature, they create the next Golgi compartment. Intriguing new data suggest that perhaps neither model is completely correct. This will likely lead to yet another model. You may not see what all the fuss is about, but the differing Golgi theories say very different things about how cells function. Understanding basic cellular processes, such as how the Golgi works, ultimately can have a profound impact on the development of methods to diagnose, treat, and prevent diseases that involve those processes.

6

Micro Propagation of Plants

Tissue culture techniques have been used to help produce crops in large scale through mikropropagasi or klonal propagation of various plants. Plant tissue in very small amounts can produce hundreds or thousands of plants continuously. This technique has been used in industrial scale in various countries to commercially produce various types of plants such as ornamental plants (orchids, cut flowers, etc..), Fruit crops (like bananas), crops and forestry industries (coffee, teak, etc.). By using tissue culture methods, millions of plants with the same genetic characteristics can be obtained only from one eye buds. Therefore, this method becomes an alternative in the vegetative propagation of plants.

IMPROVED CROP

In crop improvement efforts through the glorification of the conventional methods, to obtain pure strains can take six to seven generations of self-pollination or crosses. Through tissue culture techniques, can be obtained homosigot plants in a short time by producing haploid plants through pollen culture, anther or ovaries followed by chromosome doubling. Homosigot plants can be used as plant breeding material in order to improve the nature of the plant.

PRODUCTION OF DISEASE-FREE PLANTS (VIRUS)

Tissue culture technology has contributed in a plant that is free from viruses. In plants that have been infected with the virus, the cells in the bud tip (meristem) is an area that is not infected with the virus. In this way the meristem will mengkulturkan obtained virus-free plants.

GENETIC TRANSFORMATION

Tissue culture techniques have become an important part in helping the success of plant genetic engineering (gene transfer). For example, bacterial gene transfer (such as cry genes from Bacillus thuringiensis) into the plant cells will be expressed after transgeniknya achieved plant regeneration. The production of secondary metabolites, compounds Plant cell culture can also be used to produce biochemical compounds (secondary metabolites) such as alkaloids, terpenoids, phenyl etc. propanoid. This technology is now available in industrial scale. For example, the commercial production of compounds "shikonin" from Lithospermum erythrorhizon cell culture. Plant cells can be grown in isolation from intact plants in tissue culture systems. The cells have the characteristics of callus cells, rather than other plant cell types. These are the cells that appear on cut surfaces when a plant is wounded and which gradually cover and seal the damaged area. Pieces of plant tissue will slowly divide and grow into a colourless mass of cells if they are kept in special conditions.

These are:

- Initiated from the most appropriate plant tissue for the particular plant variety
- Presence of a high concentration of auxin and cytokinin growth regulators in the growth media
- A growth medium containing organic and inorganic compounds to sustain the cells
- Aseptic conditions during culture to exclude competition from microorganisms

The plant cells can grow on a solid surface as friable, pale-brown lumps (called callus), or as individual or small clusters of cells in a liquid medium called a suspension culture. These cells can be maintained indefinitely provided they are sub-cultured regularly into fresh growth medium.

Tissue culture cells generally lack the distinctive features of most plant cells. They have a small vacuole, lack chloroplasts and photosynthetic pathways and the structural or chemical features that distinguish so many cell types within the intact plant are absent. They are most similar to the undifferentiated cells found in meristematic regions which become fated to develop into each cell type as the plant grows. Tissue cultured cells can also be induced to re-differentiate into whole plants by alterations to the growth media.

Plant tissue cultures can be initiated from almost any part of a plant. The physiological state of the plant does have an influence on its response to attempts to initiate tissue culture. The parent plant must be healthy and free from obvious signs of disease or decay. The source, termed explant, may be dictated by the reason for carrying out the tissue culture. Younger tissue contains a higher proportion of actively dividing cells and is more responsive to a callus initiation programme.

The plants themselves must be actively growing, and not about to enter a period of dormancy. The exact conditions required to initiate and sustain plant cells in culture, or to regenerate intact plants from cultured cells, are different for each plant species. Each variety of a species will often have a particular set of cultural requirements. Despite all the knowledge that has been obtained about plant tissue culture during the twentieth century, these conditions have to be identified for each variety through experimentation. Plant cell culture is viewed as a potential means of producing useful plant products such that conventional agriculture, with all its attendant problems and variables, can be circumvented.

These problems include: environmental factors (drought, floods, etc.), disease, political and labour instabilities in the producing countries, uncontrollable variations in the crop quality, inability of authorities to prevent crop adulteration, losses in storage and handling. Thus, the production of useful and valuable secondary metabolites in large bioreactors located in the consuming country is an attractive proposal. Additional advantages of such processes include: controlled production according to demand and a reduced and requirement.

However, this technology is still being developed and despite the advantages outlined above, there are a variety of problems to be overcome before it can be adopted on a wide scale for the production of useful plant secondary metabolites. The success of Mitsui Petrochemical Industry Co. Ltd. in Japan in producing shikonin on a commercial scale from Lithospermum erythrorhizon cultivations and that of Nitto Denko Co. Ltd. also in Japan in mass production of Panax ginseng or ginseng cells using 20 kL tanks have demonstrated that many of the problems can be overcome with perseverance. The economic feasibility of these processes is another question and this will be dealt with in a separate section. In theory, it is anticipated that such large scale suspension cultures will be suitable for industrial production of useful plant chemicals such as pharmaceuticals and food additives, in a manner similar to that of microbial fermentation.

Nevertheless, there are some significant differences between microbial and plant cell cultures that must be considered when attempting to apply plant cell cultures to the available technology. Table 1 shows a comparison of some of the characteristics of plant and microbial cultures of relevance to fermentation. This table serves to demonstrate some of the problems that can be encountered with plant cell cultures. The sensitivity to shear is due both to the large size of the cells and to the relatively inflexible cellulose cell wall. Thus, with normal blade impellers the cells may twist which will inhibit mitoses and, for this reason, air-lift fermentors are recommended by some researchers. The large size of the plant cell contributes to its comparatively high doubling time (12 h - several days), which thus prolongs the time required for a successful fermentation run.

The vacuole is the major site of product accumulation, and since product secretion is uncommon, the high metabolite yields seen in microorganisms that secrete product (thereby removing product inhibition of biosynthesis) cannot be expected. There is some ongoing research on membrane permeabilization of plant cells which may serve to relieve the constraints of product inhibition by facilitation of leakage into the extracellular medium. If this would also permit recycling of the biomass (*e.g.* via immobilization) it would help reduce production costs.

The low aeration requirement for plant cells is an advantage over microbial cultures in general. In addition, the high cost of running a fermentation vessel over several weeks should be considered, although media costs are much less than those of animal cell cultures.

Table. Characteristic of Microbial and Plant Cell Relevant to Fermentation

Characteristics	Microorganism	Plant Cell
Size	2 u	>10 u
Shear stress	Insensitive	Sensitive
Water content	75%	>90%
Duplication time	<1 hour	days
Aeration	1-2 vvm	0.3 vvm
Fermentation time	Days	Weeks
Product accumulation	Medium	Vacuole
Production phase	Uncoupled	Often growth-linked
Mutation	Possible	Requires haploids
Medium cost (₹)(MS medium)	8-9/m	65-70/m

In addition to the problems outlined above, concerned with fermentation technology, there are also considerable hurdles to be overcome at the biochemical level. The two major problems concern poor expression of products and instability of cell lines.

Cultured plant cells often produce reduced quantities and different profiles of secondary metabolites when compared with the intact plant and these quantitative and qualitative features may change with time. The poor product expression is often attributed to a lack of differentiation in cultures. On the other hand, there are cases of cultures that over-produce metabolites compared with the whole plant.

There are a number of examples of cultured cells producing metabolites not observed in the plant, eg. Lithospermum erythrorhizon cultures have been observed to synthesize rosmarinic acid. It has become apparent that the choice of original plant material having high yields of the desired phytochemical may be important in establishing high-yielding cultures.

Furthermore, the need to repeatedly screen for high-producing lines (due to inherent instability of cell lines) has been emphasized, although the nutritional composition of the medium is also important. Thus, a variety of approaches are being investigated by many researchers to increase productivity of useful plant metabolites in plant cell cultures as seen in another section in this review.

Table. Secondary Metabolites Produced in High Levels by Plant Cell Cultures

Compound	Plant Species	Yields	(% Dry Wt)	Culturetype*
			Culture	Plant
Shikonin	Lithospermum erythrorhizon	20	1.5	s
Ginsenoside	Panax ginseng	27	4.5	c
Anthraquinones	Morinda citrifolia	18	0.3	s
Ajmalicine	Catharanthus roseus	1.0	0.3	s
Rosmarinic acid	Coleus blumeii	15	3	s
Ubiquinone-10	Nicotiana tabacum	0.036	0.003	s
Diosgenin	Dioscorea deltoides	2	2	s
Benzylisoquinoline Alkaloids	Coptis japonica	11	5 - 10	s
Berberine	Thalictrum minor	10	0.01	s
Berberine	Coptis japonica	10	2 - 4	s
Anthraquinones	Galium verum	5.4	1.2	s
Anthraquinones	Galium aparine	3.8	0.2	s
Nicotine	Nicotiana tabacum	3.4	2.0	c
Bisoclaurine	Stephania cepharantha	2.3	0.8	s
Tripdiolide	Tripteryqium wilfordii	0.05	0.001	s

* s = suspension; c = callus®

As seen in many reviews, studies on the production of plant metabolites by callus and cell suspension cultures have been carried out on an increasing scale since the end of the 1950's. The large scale cultivation of tobacco and a variety of vegetable cells was examined from the late 1950's to early 1960's by Tulecke and Nickell at Pfizer Inc., Mandels et at. at the Natick Laboratories in the U.S. Army, Street et al. at the University of Leicester and Martin et al. at the National Research Council of Canada. Their results stimulated more recent studies on the industrial application of plant cell culture in many countries.

Since Japan has a highly developed fermentation technology, many industrial companies, in collaboration with some university groups, have tried to apply this technology for the commercial production of useful compounds. The Japan Tobacco Inc.'s interest involved around mass-production of tobacco cells as raw materials of cigarettes; the company established 20 kL fermentors which were the largest for plant cells in 1970's. Meiji Seika in Japan also elucidated the fundamentals of production of Panax ginseng cells in large volumes. Researchers reported that cultured ginseng cells stimulated physiological activities in animals in a similar fashion as elicited by native ginseng roots. The work was followed by Nitto Denko Co. which has been manufacturing cell mass of ginseng commercially. The cells are used as health foods in Japan. Researchers of Kyowa Hakko conducted extensive pharmacological screening of numerous cell cultures and found various novel products of great interest, including plasmin inhibitory proteins in Scopolia japonica cells and plant virus inhibitors in cultured cells of Phytolacca americana and other species.

The virus inhibitors of P. americana are now being studied by many research groups in the world because of their activity against AIDS and other animal viruses. Other firms such as Ajinomoto and Nippon Shin-yaku also made efforts to increase the level of accumulation of alkaloids, steroids and other secondary products in cultured cells. Groups in Germany outlined very interesting approaches to industrial application in the meeting held in 1976 at Munich, and their excellent results encouraged researchers in other countries.

For example, Zenk and his colleagues presented an impressive paper in which they successfully selected high-alkaloid producing cell lines of Catharanthus roseus using a method similar to the microbial mono-colony isolation technique. A number of laboratories in industries and universities followed Zenk's approach, and in fact, some researchers could increase significantly the level of secondary products such as ubiquinone-10, biotin, and various plant pigments produced by cell cultures. In 1982, the 5th International Congress of Plant Tissue and Cell Cultures was held in Japan and about papers presented there related to production of secondary metabolites in cultured cells and several papers seemed to be commercially promising such as production of shikonin by Fujita et al.) of Mitsui Petrochemical and that of several antitumor compounds by Misawa et al.

At subsequent International Congress of Plant Tissue and Cell Cultures in Minneapolis in 1986 and that in Amsterdam in 1990 as well as other meetings such as the Meeting of Primary and Secondary Metabolism of Plant Cell Cultures held in Giessen, Germany and the 4th and International Congress on Phytotherapy held in Munich, Germany in September, 1992, many compounds were shown to be accumulated by plant cell cultures and many different strategies were presented to increase their productivity. Means for production of those compounds include not only de novo synthesis but also biotransformation processes.

A biotransformation process to produce ß-methyl digoxin using Digitalis lanata cells studied by Reinhard and Alfermann in Germany was evaluated by Boehringer Mannheim Co. using 4 kL bioreactors although it has not yet been commercialized. A combination of a plant cell culture process and a simple chemical coupling reaction was invented by a Canadian company,

Allelix, to manufacture vinblastine as a commercially feasible process. The technology is now being studied by Mitsui Petrochemical in Japan for commercialization. In spite of remarkable advances in plant cell culture technology, the production cost of metabolites is still high, as estimated by Zenk and by Goldstein.

The former calculated that the cost would be U.S. $500 per kg of isolated product when it was accumulated in 1 g/L within a period of 15 days in 100 m batch culture. It is true that the cost has decreased since his estimation in 1974, and the producing ability of 1 g/L for 15 days is achievable in the case of some compounds such as rosmarinic acid, but it is still too expensive to produce food stuffs, food additives or pharmaceuticals which can be more easily produced by alternative ways such as chemical synthesis, extraction from plants or by microbial fermentation.

Zenk stated that industrial plant cell culture techniques would be introduced only if the plant product under consideration is produced at a price equal to or preferably lower than the field-produced product. The only factor determining the industrial realization of plant cell culture is the price by which a given product can be produced.

A very detailed assessment of the costs entailed in theoretical processes has been made by Goldstein et al.

Four cases were considered:

A	*Base state of technology*: Final biomass concentration = 200 g fresh wt./L One day doubling time, biomass = 0.693/day One day doubling time, product = 0.693/day Product concentration = 0.05% fresh weight Recycled biomass = 50%
B	*As for A, but*: One day doubling time, product = 6.93/day (10 times increase) Recycled biomass = 90%
C	*As for B, but*: Final biomass concentration = 500 g fresh wt./L (2.5 times increase)
D	*As for C, but*: Product concentration = 0.5% fresh weight (10 times increase).

The final case (D) would translate into a production of 17.5 g product/L/day, which is well in excess of levels achieved with current technology. The rate of product formation is the major obstacle, but this could be compensated for by increased concentration of product. The calculated manufacturing costs and profit margin for each case assuming production of 10^4, 10^5 or 10^6 kg/yr are shown in Tables. Note that these figures reflect 1980 prices and will have increased since then.

Table. Manufacturing Costs for Cases A - D

Production Level	104	105	106
(kg/yr)	Cost ($/kg)		
Case:			
A.	551	241	106
B.	323	142	61
C.	165	72	32
D.	92	40	18

Table. Profitability of Cases A - D. The Selling Price ($) and Profit (%) are given for each of the three Annual Production levels.

Production Level (kg/yr)	104		105		106	
Case:	$	%	$	%	$	%
A.	1,608	5.7	683	14.7	290	13.8
B.	992	17.5	421	16.3	177	15.3
C.	528	18.7	223	17.6	94	16.6
D.	233	10.6	100	9.8	43	9.1

The selling price was calculated to take into account a 5-year payment on capital and a tax rate of 50%. The formula used was: Selling price/kg = 2 x manufacturing cost/kg + 0.4 (total invested capital/kg) = depreciation/kg. In the case of scenario (D) with production of 10 kg/yr the selling price could be as low as ₹ 43/kg and the manufacturing cost would only be Rs18/kg. This assessment makes the future of commercial plant cell culture fermentation appear quite viable, but the crucial problems to be overcome are enhancing product levels, recycling the biomass (*e.g.* through immobilization) and raising the biomass concentration.

Some encouraging advances in recent years suggest that these problems may be solved eventually, *e.g.* shikonin can be produced at a level of 23% of the dry weight (roughly 2.3% or more of fresh weight) and 4 g/L; rosmarinic acid can be produced at a rate of 1 g/L/day in Coleus blumei with a final concentration of 27% of the dry weight. A biomass concentration of 40 g dry wt./L was achieved.

Batch fermentation processes are capital intensive, requiring a high initial investment and sophisticated technology. In comparison with microbial cultures, doubling times are long and product yields are low. Furthermore, since production is often not growth-linked, there are many cases where two stage production is necessary, thus increasing costs markedly. Table 5 shows estimated costs for a hypothetical phytochemical produced by plant cell culture fermentation.

Table. Batch Fermentation Costs for a Hypothetical Plant Product as a Function of Product Concentration

Product Concentration (%)	0.1	1.0	10.0
Fixed Capital Investment ($M)	340	84	21
Total Production Cost (%/kg)	5,900	1,045	228
Assumptions:	Annual production = 20,000 kg Cell doubling time = 60 hr Batch cycle time = 15 d Cell intensity = 20 g/L		

The total production cost is the sum of direct and indirect costs, including raw material, labour, utilities, capital eq;uipment depreciation over a 10 year period, maintenance, tax and insurance. Indirect costs were taken as 15% of capital investment It has been calculated that if rose oil were produced in cell cultures, a fermentation process with a 20% share of the market would yield a net profit of $0.5 M/yr. This assumes a rose oil concentration of 10% dry weight and a cell density of 20 g dry weight/L. Fowler suggests that plant products valued at $250-1,500/kg are acceptable production targets for an economically viable process. An initial market penetration of the world or U.S. market has been recommended at between 10 and 20%.

This will depend on the size of the market and whether a plant cell fermentation process opens up new market. Scragg has calculated the effect of run time on productivity. Assuming a product yield of 1% dry weight, biomass yield of 20 g dry weight/L and a bioreactor volume of 1,000 L. A 300 day year was estimated, since time would be required for maintenance, contaminated runs, etc. Figure shows that as productivity decreases the requisite run time increases. Similarly, as the yield of product increases the reactor volume required decreases expotentially. Under the conditions described above, a 100,000 L fermentor would be necessary to produce 300 kg/yr.

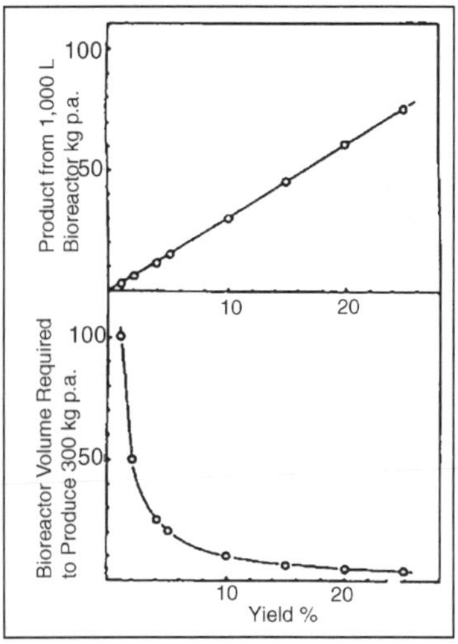

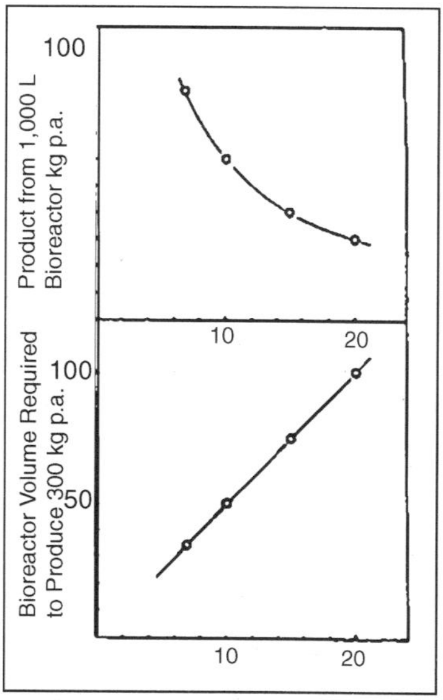

Fig. The Effect of Yield on Bioreactor Volume and the production from a 1,000 L. Bioreactor

The most detailed assessment available in the literature for the production of a specific product in plant cell cultures is concerned with the production of ajmalicine from Catharanthus roseus cell cultures. In this case, a 20% market penetration was assumed, ie. 800 kg/yr and production was based on the use of a two-stage batch culture process. Stage two parametres include specific productivity of 0.26 mg/g/day (based on final dry weight), final product concentration of 0.06% dry weight and a maximum fresh weight concentration of 160 g/L. It was estimated that the cost of production would be Rs3,215/kg ajmalicine (Rs7.30/lb dry biomass) and this compares with Rs619/kg (Rs0.70/lb dry biomass) from the intact plant. Clearly, this 5-fold increase in costs is unacceptable from a commercial point of view.

This example serves to illustrate the need to chose products that are particularly expensive to produce in the field. Shikonin is a good example; in that the long growing period and strict climatic requirements mean that the cost of the plant raw material is high \$6.80/lb. It has been suggested that a quick way to assess the attractiveness of a cell culture method vs. conventional agriculture, is to calculate a specific biosynthesis rate "based on the final dry weight and the total time of fermentation or land occupation". With shikonin there is a 830-fold increase in plant cell culture, whilst with ajmalicine from C. roseus there was only a 24-fold improvement.

CLONAL PROPAGATION

Tissue culture is well suited for quick vegetative propagation of plant species. It is used for asexual propagation in many species of fruit and timber trees and also used for obtaining disease free and virus-free plants. The major difficulty in the use of this technique in clonal multiplication is the occurrence of genetic variation among the regenerated plants. This problem can be reduced to a large extent by using young tissue cultures, preferably during the first few subcultures.

CLONING

Genetically identical plants derived from an individual are called clones. Processes that produce clones can be put under the term 'cloning'. This includes all the methods of vegetative propagation such as cutting, layering, and grafting. Propagation by tissue culture also helps in producing clones. Using the shoot tip, it is possible to obtain a large number of plantlets. This technique is used extensively in the commercial field for micropropagation of ornamental plants like chrysanthemum, gladiolus, etc. and also crops such as sugar cane, tapioca, and potato. Thus an unlimited number of plants that are genetically similar or are clones can be produced in a short span of time by tissue culture.

CLONAL PROPAGATION OR MICROPROPAGATION

It has been demonstrated that a variety of plant species can be conveniently propagated through the techniques of cell, tissue or organ culture. This is popularly described as clonal propagation ormicropropagation.

The major benefits of this method include the following:

- Rapid multiplication of superior clones and maintenance of uniformity;
- Multiplication of disease free plants
- Multiplication of sexually derived sterile hybrids.

In most cases, clonal propagation is achieved by placing sterilized shoot tips or auxillary buds onto a culture medium that is sufficient to induce formation of multiple buds.

This method has already been used to propagate a large number of marketable ornamentals.

Following stages are involved in the method of clonal propagation:

- Stage I involves establishment of tissue in vitro.
- Stage II involves multiplication of shoots (often media are not changed between stage I and stage II).
- Stage III concerns root formation and conditioning of propagules prior to transfer to the green house; this stage requires alteration of media for promotion of root formation.
- Stage IV involves growth in pots followed by field trials. The number of steps in micropropagation may sometimes be reduced to three or two as shown in diagram.

Three Different Methods of Micropropagation of Plant Material

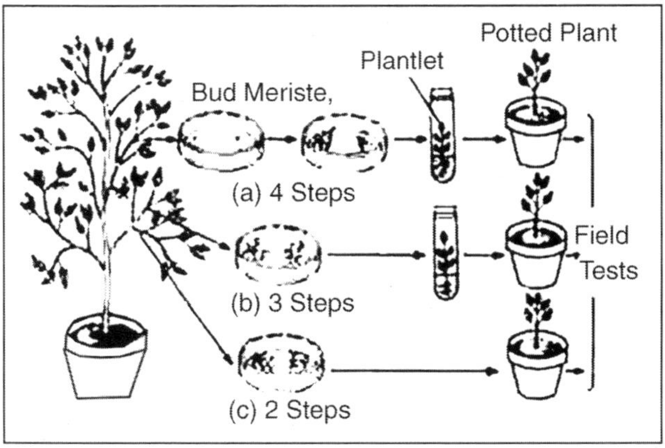

A wide range of plants have now been regenerated through technique of tissue culture. This has been found particularly useful for propagation of tree species, so that a large number of plant species have been successfully grown by tissue culture.

LARGE-SCALE PROPAGATION

To bridge the gap between research and application, the Department of Biotechnology, Government of India sponsored the setting-up of two pilot-scale facilities for large-scale propagation of elite planting material of forest trees through tissue culture. One of these facilities has been established at TERI's 36-hectare-campus in Gual Pahari, Haryana with an annual capacity of a million plantlets. Research at these facilities focuses exclusively on developing new protocols for mass cloning of elite planting material, mainly of trees.

Till date, over 4 million plants have been dispatched for field plantation from these facilities. The tissue culture raised plants are presently being evaluated under field conditions. This is being done in tandem with the forest departments of Haryana, Uttar Pradesh, Madhya Pradesh, Bihar, Jammu and Kashmir, and Orissa. For initial screening for phenotypically superior trees only a few hundred plantlets of the same are raised and tested under various agroclimatic zones. The best clones are then mass multiplied and monitored regularly for their performance. Field data suggest a survival percentage of more than 90% even in the harsh conditions of Aravalis without the life-saving irrigation. At half the rotation age some of the selected clones of Eucalyptus are showing a significant increase in productivity as compared to the conventional seed raised progenies.

ADVANTAGES OF CLONAL PROPAGATION

Clonal forestry is often seen as modern biotechnology. It is true that some techniques used for creating clones are new. The technical capacity to produce clones is improving. The current paper aims to review the possible advantages of clones in forestry and forest tree breeding and to give information about the occurrence of clones in nature and forestry. Some specific attention is given to focus on matters, which are relevant for cloning as a tool to get quality birch. No effort is done to discuss possible disadvantages and problems in full, but this has been done many times before. Neither are the possibilities of other "competing" alternatives to get some of the advantages (like using controlled crosses) fully explored.

Clones in Nature and Forestry

Clones are common in Nature, and they have been used as a tool of domestication for very long time and are used in large-scale forest operations. It seems to be important to remember that to put clonal forestry into perspective. The earliest land plants lacked vascular tissue and had inefficient systems of sexual reproduction and dispersal and clonal reproduction was the typical way of propagation in nature. Clonal growth was advantageous enabling physical dominance of large areas through horizontal growth.

Clonal propagation is still important in nature for many species. Many of the commonest clones are herbaceous but the habit is also found in woody plants throughout the world. In temperate broad-leaved woodlands, root suckering is a common form of clonal growth and there are good examples in the genera Populus and Prunus. Coppice is a natural growth form that has been exploited for centuries in European forests. Betula, Carpinus, Corylus, Quercus, Salix and Tilia are all genera with the ability to self-coppice.

The use of vegetative propagation of trees as a tool in their domestication has a long history. An early use was to propagate good clones of fruit trees and this has been done for thousands of years, thus humanity has long experiences of cloning trees.

For specific cases clonal forestry is often regarded as a "new technique", and "biotech" leads associations to the public to something more radical than most cases of clonal forestry. To put things into perspective it has to be remembered that many of the techniques used in modern "seedling forestry" are also "new", and the problems and uncertainties with that is often larger than the uncertainties connected to clones.

Sugi

The first known use of vegetative propagation for forestry purposes was in Japan with Sugi in the 15th century. Since then reforestation with monoclones or clonal mixtures has been widely used in Japan. Cryptomeria japonica is the most widely cultivated tree species in Japan comprising nearly half of the ten million ha of Japanese plantations. Almost half of the plants used are clones.

The extreme uniformity of the plantations has probably contributed to greater than average damage from typhoons and heavy snow. Attacks from Sugi bark borer and bark midge have increased in severity as the area of Sugi plantation has grown, but there is no evidence of altered disease or insect virulence related to clonal forestry, despite careful monitoring.

Clonal forestry is most common on the south-western island of Kyushu where about 100 cultivars are in use. Many foresters use only 1-3 cultivars of local origin in their plantations so that a single clone may cover as much as 2 hectares. Allergy to Sugi pollen is a health problem in Japan as a whole, but is only a minor problem on Kyushu with its large use of clones. Cloning by cuttings for several generations has reduced the incidence of male flowers, so these plantations produce little pollen. Analogously one may speculate in that a reversion of North European birch forestry to clonal forestry in a very remote future may reduce allergic reactions with birch pollen.

Eucalypts

Clonal forestry is more common with broadleaf species than with conifers. Probably the most successful example is the eucalypt plantations in countries in tropical and subtropical climates like South America, South Africa and Portugal. The total area of eucalypt plantations today is about 30 million hectares and about half of this area is planted with cloned material.

The eucalypt plantations in Brazil cover an area of approximately 3 million hectares. A succession of eucalypt species has been grown in Brazil due primarily to disease problems. Clonal forestry started in the late 1960:s and is today the dominant form of eucalypt forestry in the country. Progeny from an Eucalyptus grandis female (male sterile) growing in an arboretum were found to be natural hybrids which were resistant to pests and diseases. The offspring of this tree are believed to be Eucalyptus grandis ′ Europhylla hybrids.

This knowledge was rapidly adopted by the forest industry companies and they selected clones within the open-pollinated offspring from this single tree, which means that a small number of clones originating from the same family are widely grown over large areas in Brazil. This has and is still changing with most companies having good breeding programmes backing their clonal programmes. The clones are deployed in huge monoclonal blocks. This situation seems favourable for a pathogen, but nothing has happened yet - despite the possible risks of a disease in a monoculture.

Today the companies have become more risk-conscious and are developing breeding and clonal testing programmes that include strategies for clonal diversity in space and time. A typical programme is now deploying about 20 clones each year and these clones will be replaced with time as new clones are selected. The production landscape forms a genetic mosaic. New genetic material has been imported to widen the genetic base for breeding.

Poplars and Willows in Europe

Vegetative propagation of poplars (Populus sp.) has a long history in Europe. Organised clonal forestry started in the beginning of the 20th century. Monoclonal plantations of poplars have become a common land use especially on river plains in southern European countries like Italy, Spain and France. Some of the best poplar clones in use were selected as early as during the first half of the 20th century. Clonal forestry with poplars is common in countries with subtropical and temperate climates like Italy, Spain, France, Belgium, USA and Canada. Single clones of poplars have been propagated extensively. There have been disease problems and the overuse of single clones and use of large monoclonal plantations has been questioned.

Poplar has covered over 100000 ha in one country and a single clone may comprise one third of this area. The focus has been to develop disease resistance within individual clones, but these clones have proved to be highly susceptible to new varieties of disease. The disease problem seems to be growing and has caused Germany to more or less abandon poplars and other countries may follow.

Nevertheless these incidents have not provided poplar growers with sufficient incitement to focus on more diverse alternatives. It is unlikely that monoclonal cultures and the wide use of some clones are the only explanation to increased disease, but they act as contributing factors. Legal and commercial reasons often favour the use of single clones. Even willows and short rotation forestry in Sweden are threatened by disease. Resistance is broken down within a single rotation (it is decades in between replanting even when harvesting is done at intervals of a few years). Two evaluations of the Swedish willow breeding programme have rather strongly recommended an increased number of selections to be made (more than a single selection released per year). Nevertheless, the growers′ organisations still do not encourage clone mixtures.

Hybrid Aspen in Finland

A Finnish forest industry company has recently started a clonal forestry programme with hybrid aspen. This raises the question if something similar could be done with birch. It may be noted, however, that suggested that micro propagation costs are high and that the technique needs development.

Overoptimistic Statistics

It is a clear tendency that statistics reported about the use of vegetative propagation, at least there it is regarded as modern hitech biotech, gives the impression of more rapid progress than actually occurs. It is reported, based on a survey, that four Swedish operators produces 7.3 millions rooted Norway spruce cuttings annually. It is estimated the annual production to 5 millions and believed in a raise to 10 millions. However, looking backwards, less than

20 million were planted in total till 2000 and now only some hundred thousands are planted annually.

Crop Uniformity

As genetic diversity and possible uniformity is an evident and possible important aspect of clonal forestry. For evaluation of possible ecological impact diversity seems central. I will thus start to discuss disadvantages with a low genetic diversity in a stand. In other respects this paper focuses on the advantages of clonal forestry.

Disadvantages of a Uniform Crop

There are reasons to believe that the biological production can be higher in a genetically diverse crop than in a uniform:

- A single genotype demands the same things at the same time, and thus utilises the site worse than a mixture of genotypes.
- In a mix, another genotype may take over the ecological space left by a failed genotype.
- A disease spreads faster in a uniform crop.

These statements are supported by the average of a very large number of agricultural mixing experiments and a few experiments and experiences with forestry. On average, mixes performed some per cent better and were less susceptible and more stable than the average of its components, but usually the production of the mix was not better than its best component grown pure.

Even if mixes seem more productive on an average, there are many examples and experiments, when no advantage was found and some experiments where diversity actually seems negative for production. I conclude that from a production point of view an advantage with some degree of diversity is expected, but that does not imply that production is higher the higher the diversity is. But the superiority in biological production is not expected to be large and it is not certain for individual cases, but just a rough generalisation. This general statement ought to apply to birch also, thus biological production is assumed to be lower and more uncertain in a uniform crop, but the effect is probably small and practically unimportant (a few per cent of biological production).

Advantages of a Uniform Crop

There must be considerable advantages in uniform monoclonal blocks as they are used frequently. Although the advantages with diversity are repeated and reiterated and very politically correct, it is seldom applied in intensive agriculture or husbandry. The main reason is that maximizing biological production and maximizing economical production is not the same thing. It makes it easier to manage the crop if it behaves uniformly. For a forester it may not matter very much if individual birches drop their leaves on different days, but a farmer wants the crop to become ripe the same day, as the whole crop is harvested the same day. Diversity is a usually a problem for the customer, who wants a uniform raw material for further processing. For administrative processes, like describing the crop, it is an advantage if it is uniform and does not change from specification over time and space. These problems are evident for many types of agricultural crops, and therefore these crops are usually uniform even when biological production would be higher with a diverse crop.

The dream of many foresters is to find the best genotype and use only that. This has been successfully done for other domesticated species. The uniformity itself in a clonal plantation has a considerable advantage from a management point of view. It is easier in the nursery to raise plants of a single genotype. It is easier to manage and harvest a uniform forest. It has advantages to handle a single type of material. A customer or processor of a product knows more about a lot if it is a specified clone than if just the species is known. A batch of uniform logs is worth more and easier to handle than the same amount of very variable logs.

Even if interest is only in quantitative production and that is lower in a uniform crop, still the higher genetic performance of the best clones compared to the best seedlings available usually much more than compensates for the production loss by a uniform crop. Thus uniform clonal crops produce more than diverse seedling crops.

My opinion is that genetic diversity often is overemphasised in comparison with the higher genetic gain achievable at the cost of a reduction in genetic diversity for future forest production plantations in many countries (like Sweden) both for propagation with seedlings and cuttings. To get genetic diversity many clones are used in seed orchards and clonal mixtures and there constraints of relatedness of clones may be seen as desirable. This was not a disadvantage at an early stage of forest tree improvement, as it does not influence gain much if many phenotypically selected plus trees are placed in seed orchards, and in the early stages of cutting propagation there were physical constraints on the number of ramets which can be produced by each clone, so it is necessary to have many clones. But now this constraint often means selection is not intensive and the best genotypes are not exploited to the extent, which would be possible.

However, clonal forestry is unlikely to deal with a single clone only. That does not seem a likely scenario for birch. Many clones will be tested and test results as well as experiences will accumulate of the best clones, so the opinion of what is the best clone will change over time. It is also dependent on the locality, the desires of the user and the availability. Even if uniform plantations are planted, there will usually be a mosaic of clones on the landscape level. From the risk management point of view a clonal deployment philosophy offers advantages. If a clone is sensitive to a pest or pathogen the problem can be identified and managed. If silvicultural means to deal with a problem are insufficient, the stand can be salvage harvested and replanted. The clone which turns out inferior can quickly be taken out of production. It is an advantage for an operator which uses monoclone blocks to use several clones in the same time that it will become more evident what the strong and weak point of the clones are.

A disadvantage with mosaics is that the uniform lots of wood from a stand are often too small to be practical as a specific variety at the end user, and actually the end user may get trouble with larger between lot differences if clonal forestry is used. Alternatively the clones can be grown in intimate mixtures, in that they most of the possible advantages of the diversity of seedlings may be gained by a mix of rather few clones.

When we are talking quality birch - and not quantity birch - the reasons for monoclonal plantings become stronger. When high quality of the product is concerned, it often becomes more important to have uniform quality also, while in a low quality product - as firewood - the quantity is usually relatively more important and a uniform and predictable product is less important. Attention on quality instead of quantity also generally coincides with a higher level of intensity, and at a higher level of intensity, clonal forestry often becomes more interesting. Thus considering the aim the project arranging this symposium has in mind; uniform clonal plantings becomes relatively more important than for forestry in general.

IMPROVED CLONES CAN BE PREFERABLE TO SEEDLINGS

There are a number of arguments for forestry with clones. It takes less time to multiply a genotype as a clone than to build a seed production unit with that clone as one of the parents and wait till that works. The time lag is actually much longer than till the first seed production, because a seed production unit is usually designed to last for many years. The risk for getting the seed storage empty means that forestry often want to have five to ten years foreseen seed need in storage. That is expensive, many seeds may never be used and the seeds actually used are less improved than possible. A seed production unit is designed to meet a certain need of plants with certain characteristics many years ahead.

If too many seeds are produced it means that the investment was unnecessary large. If too few seeds are produced it may mean that foresters must use genetically inferior unimproved seeds. If the propagation is done fast by vegetative propagation by *e.g.* micro propagated clones or somatic embryogenesis, it takes only a year to multiply a desirable genotype in the desired number of copies. Thus it is fairly easy to adapt to the latest news from breeders and customers. There are large annual variations in seed production both in the forest and in seed orchards. In a seed orchard the reproductive output is variable among genotypes and the reproductive phenology may make clones incompatible, so the actual genetic set up of the progeny is a bit uncertain. These uncertainties can be avoided by clonal forestry. Many of our forest trees - like birch - can both spread pollen and produce seeds. That means that selfing can occur and the occurrence of selfing in a plant material will decrease the forest production. With the use of clonal forestry selfing can be completely eliminated.

In a wind pollinated seed orchard wild uncontrolled and unimproved pollen may contribute to pollination, clonal forestry is a way to eliminate this problem. However, for birch, seed orchards are small enough to be managed in plastic green houses, where the inflow of unimproved pollen is low, so for birch it is not fair to see this pollen contamination problem as an argument against open pollinated seed orchards. In a seed orchard it is important to avoid related clones, as relatives get inbred progeny when they mate, which means that forest production drops because of inbreeding depression. Mating between relatives may be a worse problem than selfing in future seed orchards, as self-zygotes usually die; self-seedlings are often culled before planting; or early out-competed in forest, while milder inbreeding is more directly transferred to production loss. If clones are related in a clonal forest, that does not cause inbreeding depression, and thus there is little harm if related clones occur in a stand (actually this is very natural, trees in a natural stands are often related). The clones in the breeding population will be increasingly more related as breeding goes on, it will be easier to handle the consequences of this for the production forest with clonal forestry.

A seed orchard is a big investment intended to meet the seed need for a long period for a specific species used in a specified area, thus a considerable local market is needed to justify the investment. Conifer seed orchards are usually designed for a need of many millions plants per year over some decades. Multiplication of a clone can be seen as a smaller investment during a shorter period, thus the local market is allowed to be much smaller with clonal forestry, *e.g.* micro propagation for curly birch in Finland serves a market of 200 000 seedlings a year. A seed orchard has a stiff structure, its composition can be only slightly modified once it is established and it is not easy to justify several orchards heading for different characteristics for the same area. Clonal propagation is more flexible; an individual customer can choose the clones, which fits the particular needs for the customer and occasion. Sometimes a combination of rather rare characters is desired, a clone can be found which combines several desirable characters in a much more efficient way than dealing with seed orchards.

It is often stated that clonal forestry reduces genetic diversity. This is a generalisation, which need not always be true. Actually, clonal forestry provides a tool to choose genetic diversity at will, while when a seedling material is used, the genetic diversity is essentially left to chance. Probably those using intensively managed birch plantation for producing high quality timber with special characteristics would choose the option to reduce or eliminate genetic diversity on the stand level, because they found it economically sound to do that. It seems likely this is how the tool will be used. But clones in a clonal mix can be chosen so that the genetic diversity is larger than in a seedling material.

Clones can be mixed with seedlings, that may be a good strategy not only for diversity, but economically in cases where cloned plants are much better, but also much more expensive. If the clones succeed well, their share at the final harvest will be larger than at establishment. Over time and space different clones will be used even in a system with monoclone cultures (like Eucalypts in Brazil). In such a system the genetic diversity among stands is likely to be much than in nature. Sometimes clones have been used for natural conservation in situation when good seeds could not be obtained. Clonal forestry may strengthen the trend to produce the needed wood at a limited acreage of intensively managed production forests, allowing larger areas to be set aside for other purposes. Thus the question how clones may affect diversity has many aspects and it is an oversimplification to state that it must be severely reduced. A common situation is that some good seeds exists or can be obtained, but quantitatively insufficient for the plant need. If such a situation occurs unexpected and unforeseen, vegetative propagation may be the solution. A more common situation is that a few superior seeds can be produced by controlled crosses, but too few to meet the plant demand. Clonal propagation is often a way to make forestry with controlled crosses practical feasible. Clonal propagation may be seen as a way to amplify the impact of a few good parents or a tested cross by mating the parents and when multiply the seeds.

The genetic value of a genotype depends on the breeding value of its parents, but also how well the genes of the parents fit together. This may be called dominance. The later reason for genetic differences cannot be exploited in an ordinary seed orchard, but it can be exploited with clonal forestry. Hybrids sometimes result in good seeds. It may be difficult to make many of those seeds by controlled crosses, but hybrid seeds (or trees) can be multiplied by vegetative propagation. This is used in the Eucalypts and Aspen programmes mentioned above. Personally I am not that convinced that hybrid superiority is so common as it is often stated. When hybrids are better than the parents it may be because the parents were not well adapted for the conditions where they were tested or that the hybrid combines characters of the parents for the environment and purpose of the forester in a better way than tested alternatives. Hybrid maize is a best seller, but the reason is more that farmers are unable to produce hybrid seeds them self, but has to buy them, and when it is good economy of a seed company to invest in hybrids, so they can market the seeds.

To get a genetically superior forest, it is desirable to use the genes of the very best clones. In a seed orchard it has to be rather many parents to avoid selfing and problems with lack of overlap in flowering time. This means a drop in breeding value. For clonal forestry the very best clone can be used. This is true even if in the complete absence of dominance.

As dominance is seldom the major cause of genetic variation for the most important characters in forest tree breeding, this more effective exploitation of breeding value is usually a more important reason for clonal forestry than the exploitation of dominance variation, at least as long as we do not talk about hybrids of different species or races. Quantitatively the low importance for the superiority of cloning of a limited amount of dominance variation is demonstrated.

We have also special features for clonal selection. We can intentionally select clones, which do not waist resources on sexual activities but spend resources on stem wood production. Such clones are desirable for forestry. In a seed orchard it is not desirable with clones, which do not produce seed and pollen, so such clones would not be chosen for a seed orchard, and if they were, they would be unable to transmit their genes to the seed crop.

Cloning methods have different characteristics. They are often very depending on the state of the material which is cloned, cuttings usually do not work well on physiologically old material, for somatic embryogenesis other tissue than immature seeds are often a difficult starting material. With micro propagation it is usually possible to multiply old trees or clones, and this is often an advantage.

Cloning methods have different characteristics. They are often very depending on the state of the material which is cloned, cuttings usually do not work well on physiologically old material, for somatic embryogenesis other tissue than immature seeds are often a difficult starting material. With micro propagation it is usually possible to multiply old trees or clones, and this is often an advantage.

Cloning and the cloning method can have effects on the physiological status of the tree. *E.g.* conifer grafts have not thick bark at the bottom of the trunk. The protocol for making clones can have effects on the plants. These effects sometimes look desirable and can be exploited. For example Norway spruce cuttings seems to suffer a lower mortality from pine weevil damage than seedlings. Clones usually origin from older and more mature tissues than seedlings and this affects the characters of the plants and may have effects also on the mature trees. I guess these differences to seedlings are not dangerous and nursery raised seedlings can also be claimed to be unnatural, and they my be different because of different growing regimes. I suggest that we can intentionally interfere with the characters of the plants by the production system and that clonal forestry gives additional options to use this in a positive way. Many of the factors I have mentioned above will contribute to that clonal forestry gives a genetically more improved forest than comparable seedlings. An example of the development of gain over time for the clonal option versus the seed orchard option is shown in Figure. This is based on a simulation of the Swedish Norway spruce (Picea abies K.) long term breeding programme with inputs chosen to correspond to the real programme as close as possible. Clonal forestry and seed orchards are different ways of harvesting gain from the breeding population and clonal forestry means that the genetic improvement reaches a certain level three decades before the same level can be achieved by seed orchards.

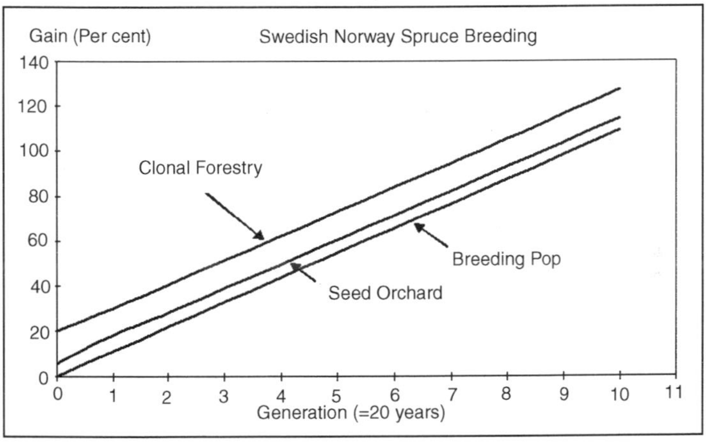

Fig. A comparison of the genetic value of three populations as a function of time based on a simulation of the Swedish Norway spruce breeding Programme.

Genetic progress is primarily created in the breeding population. The production population is created by creaming the best available material from the breeding population either by clonal forestry or seed orchards. The gain unit can roughly be interpreted as per cent gain. The details of the figure depend on exact specifications, but it gives an idea of magnitudes based on best possible estimates. Clonal forestry gives a higher gain than seed orchards; the difference at a given time is in the magnitude of 15%. The same gain is achievable from seed orchards; it just takes a few decades longer to get it.

MORE EFFECTIVE BREEDING WITH CLONES

It was pointed out that clonal forestry might be a faster way to better forests than seed orchards. Good genotypes are multiplied as grafts in most current forest seed orchards, thus clones are useful also for producing genetically improved seedlings. But there is another - still more powerful - way of using the clonal tool to improve the genetic quality of forests. Clone testing has the potential to make the long-term breeding considerable more efficient. Even if clones are too expensive for direct use in forestry, their use in testing of genetic materials may help forestry to get better seeds in the future. A combination of testing clones for forestry and for long-term breeding has synergistic effects, thus such a programme is more efficient than a programme focusing only on one of these two aspects. Higher organisms are diploid, thus have two set of homologous chromosomes and carry two packages of genes. The parents only transfer one package (a haploid gamete) to the progeny, and the composition of that gamete is different for each progeny. Thus, if offspring is used to test parents, only half of the genes in the offspring are relevant for the parent, and these genes will be different for each offspring individual. In contrast, in clonal testing all genes in the genotype are tested and no genes generate random noise in the statistical sense.

The performance of a clone depends on its father, its mother and how the genes of the mother and father match together, thus dominance effects. However, dominance variation does not seem large enough for most characters of practical interest to be important. Thus the performance of a clone in a clone test says much about what will be inherited to its progeny. The accuracy by which the breeding value of a clone is determined increases with the number of ramets.

The uncertainty caused by the environment decreases then environment is repeated. Thus clonal testing is a way of increasing the accuracy. One can see it as a way of increasing the heritability to decrease the environmental variance by letting the environment be the average of a number of individual environments, one for each ramet. The gain obtained by a genetic selection depends on the selection intensity and the accuracy of the test. Under the constraint of a fixed plant number, a lower number of genotypes can be tested if genotypes are replicated by cloning, and thus the selection intensity will decrease It thus becomes an optimization problem. However, if the individual heritability is not very high and the resource budget is not extremely tight, the optimum is to clone in at least some copies.

A seedling material is genetically diverse; the best genotypes grown pure are much better than the average of a seedling material. The performance of a seedling is mainly depending on its environment and not so much of its genes. If it is cloned, the average over the clone is more depending on the genes. Testing by clones can be seen as a way to eliminate the statistical noise. The heritability of clonal values will be much higher than the heritability of seedling values.

Each seedling is genetically unique; a scientific principle is to use replications when testing. Thus a seedling is a rather unsatisfactory material for genetic testing and conclusions based on seedlings will be dependent on model assumptions, which are uncertain and affected by statistical noise. Clone testing is less depending on model assumptions and a more direct measure and thus offers theoretical advantages! In a series of investigations the candidates for the breeding population by clonal testing instead of phenotypic testing or progeny testing, if that is at all possible.

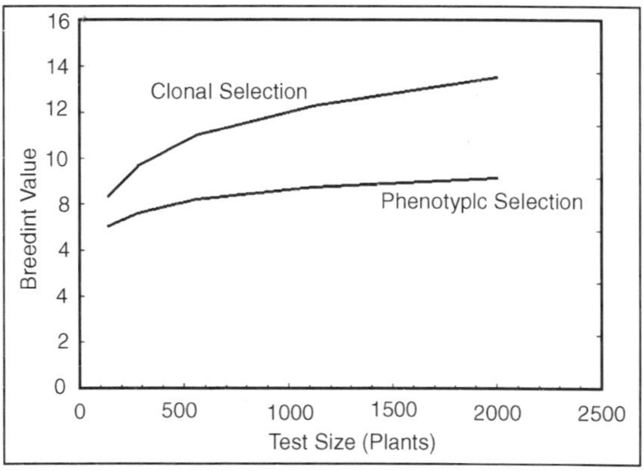

Fig. Comparison of Clonal or Seedling based Testing for the Swedish Norway Spruce long term Breeding programme (the X-axes gives the Resource level, the test Size for the real Programme is Around 560 plants/parent).

Clonal testing adds around 30% to the gain of the breeding programme, the advantage is larger the more resources that are available for each parent.

The superiority of clonal testing over phenotypic testing is demonstrated in Figure. It is based on the same type of simulations of the Swedish Norway spruce long term breeding programme as the previous figure, although it varies the resources available. The resources available today correspond to 560 test plants per parent. The more resources available, the more reason to use clone testing to get a more accurate selection (compared to increasing the number of test genotypes for a more intense selection).

Lstiburek and Danusevicius and Lindgren studied the conditions under which clonal testing is a favourable alternative. The results show that clone test is superior to phenotype test and also progeny test under a wide range of conditions. Thus even if the genotypes or ramets are expensive; and over a wide range of available resources; and in the presence of typical dominance variation; clonal testing can be recommended. In cases where the individual heritability is close to 1, clonal testing is less favourable. When the phenotype itself gives sufficient information about the genotype, the accuracy need not be improved by cloning. Such cases are rare for growth characters, but may be more common for some quality characters.

Numerous studies have shown that environment can cause lasting changes in phenotype that can be passed from one generation to the next, much as genes are transmitted. Such affects may limit the usefulness of clonal testing for breeding value, but similar objections could be made against progeny testing, and the effects must anyway be larger than usually observed (*e.g*. Lindgren and Wei 1994) to really matter.

Natural genotypes are not designed for perfect cooperation; the Darwinian selection in the struggle for life unavoidable gives some preference to strong competitors. A certain degree of selfishness is an unavoidable consequence of natural selection. By clonal selection we ought to be able to identify and utilise the best collaborators, who act in the interest of the stand and not only the individual tree..

A forest material is used over varying conditions in time and space and the forester wants a reasonable stable material over these variable conditions. Clones can be tested over time and space. That make breeders able to select clones, which perform well everywhere and anytime. Natural selection works just here and now. Again a case where Man may find more effective ways than Nature has used.

Tree Breeders first select based on field tests and when mate the selections. Artificial crosses are difficult to organise in the remote places where field tests are localised. But if clones are tested, some copies could be placed where it is convenient to make artificial mating, like in archives outside the institute. Foresters do not want trees to waist resources on sex, but to do breeding it is necessary to get reproduction to work.

But clones, which have little reproductive output in the field, can have copies at the research station, which are manipulated to make mating possible. Thus with clones different individuals can be used for test and reproduction. Genetically modified trees will play a role in practical forestry and tree breeding at some time in the future. Such implementation is much easier, if clonal forestry is an established praxis.

MUTANT ISOLATION

Biochemical mutants are far more easily isolated from cell cultures than from whole plant populations. This is because a large number of cells, 10^6-10^9, can be easily and effectively screened for biochemical mutant cells. Biochemical mutants could be selected for disease resistance, improvement of nutritional quality, adaptation of plants to stress conditions, *e.g.* saline soils, and to increase the biosynthesis of plant products used for medicinal or industrial purposes.

SOMACLONAL VARIATION

Plants regenerated from tissue and cell cultures show heritable variation for both qualitative and quantitative traits; such a variation is known as somaclonal variation. Somalconal variation has been described in sugarcane, potato, tomato etc. Some variants are obtained in homozygous condition in the plants regenerated from the cells in vitro (R_0 generation), but most variants are recovered in the selfed progeny of the tissue culture-regenerated plants (R_1 generation).

Somaclonal variation most likely arises as a result of chromosome structural changes, *e.g.*, small deletions and duplications, gene mutations, plasma gene mutations, mitotic crossing over and possibly, transposons. Somaclonal variation may be profitably utilized in crop improvement since it reduces the time required for releasing the new variety by at least two years as compared to mutation breeding and by three years in comparison to back cross method of gene transfer.

Amino Acid Analogue Resistant Mutants

Cereal grains are deficient in lysine; maize (Zea maize) is also deficient in trytophan, while wheat (T.aestivum) and rice (O.sativa) are deficient in threonine. Pulses are deficient in methionine and trytophan. Amino acid analogue-resistant cells may be expected to show a relatively higher concentration of that particular amino acid.

For *e.g.*, carrot (D.carota) and tobacco (N.tabacum) cell lines resistant to trytophan analogue 5-methyl trytophan show a 10-27-fold increase in the level of trytophan. Similarly, rice cells resistant to lysine analogue 5-(B-aminoethyl)-cysteine, show much higher levels of lysine. This technique may prove useful in the development of crop varieties with a better-balanced amino acid content.

Disease Resistant Mutants

Many pathogenic bacteria produce toxins that ae toxic to plant cells. Plant cell cultures may be exposed to lethal concentrations of these toxins and resistant clones isolated. Plants regenerated from these resistant clones would be resistant to the disease producing pathogen. This technique should be applicable to all the pathogens, which produce the disease through the action of toxin. An *e.g.*, an application is in the case of wildfire disease of tobacco (N.tabacum) produced by Pseudomonos tabaci.

Tobacco cells resistant to methionine sulfoximine, which is similar to the toxin produced by the pathogen, were isolated. Plants regenerated from these clones were resistant to wildfire disease, although to a somewhat lesser degree. The technique can be applied to those cases only where the disease is the result of a toxin produced by the pathogen. But many of the pathogens do not seem to produce a toxin, or the toxin does not appear to be the primary cause of the disease.

Stress Resistant And Other Mutants

Plant cells resistant to 4-5 times the normally toxic salt (NaCl) concentration have been isolated. Attempts to insolate such cells are being made. Similarly, attempts are being made to isolate clones that would produce more substances of medicinal or industrial value.

Somatic Hybridization

Protoplasts can be isolated from almost every plant species and cultured to produce callus. Protoplasts of two different species may be fused with the help of polyethylene glycol.

Genetic Transformation

There is some evidence that gene transfer may be achieved by feeding cells with DNA in case of eukaryotes, such as, Drosophila, Neurospora, cultured mammalian cells and in some plants. Genetic changes may be brought about by DNA or by radiation-killed pollen grains. This raises the possibility of genetic modification of plant cells with the help of both homologous (from the same species) and heterologous (from a different species) DNA. It is also proposed that DNA plant viruses, such as cauliflower (B.oleracea) mosic virus and potato leaf roll virus, plasmids (*e.g.*, Ti plasmid of Agrobacterium) and transposons, may be used as the carriers of genes for genetic modification of plant cells.

Organelle Transfer

In some cases, it may be desirable to transfer only organelles or the cytoplasm into a new genetic background. This may be achieved through the use of plant protoplasts. Chloroplasts have been transferred, and other organelles including nucleus may be transferred.

Germplasm Conservation

Tissue cultures may be frozen and stored in liquid nitrogen at -196^0C for long-term storage of germplasm. This would be of great value in the conservation of germplasm of those crops which normally do not produce seeds, *e.g.* root and tuber crops, or where it may not be desirable to store seeds. For freeze-preservation, the cells are cooled at a slow rate and are then transferred to liquid nitrogen for storage. Thawing of the cells must be very rapid for increased survival. A cryoprotectant, such as dimethylsufoxide (DMSO), is used to protect the cells from injury due to freezing and thawing. The technique of freeze-preservation *i.e.*, crybiology, of plant cells is still in the developing stages.

PROPAGATION FROM PREEXISTING MERISTEMS

Four methods for in vitro propagation of plants:

1. Shoot culture (enhanced auxiliary shoot proliferation) – The most frequently used
2. Node culture
3. Shoot organogenesis
4. Nonzygotic embryogenesis

Propagation for:

- Production of pathogen eradicated (PE) plants, and
- Preservation of PE germpalsm

SHOOT APICAL MERISTEMS

Define apical meristem: specialized region of little differentiation; retains embryonic capacity for unlimited division. Highly organized and genetically stable. Not connected to vascular system of the stem. *How does it obtain nutrients, then?* At the tip, get developing leaf primordium and lateral buds within the axils of subtending leaves.

IN VITRO CULTURE OF SHOOT MERISTEMS

Ball first detailed procedures for isolation plants from shoot meristem tips = Father of Micropropagation. Procedure used for fungal and bacterial pathogen eradication. Typically done by putting donor plant through thermo or chemotherapy and excising a meristem tip for explantation. (= meristem tip culture). Cytokinins repress apical dominance and are used to enhance axillary shoot production.

SHOOT AND NODE CULTURE

Shoot culture = propagation via repeated enhanced formation of axillary shoots from tips or buds, which are then subdivided into shoot tip and nodal segments for further secondary proliferation or for rooting as microcuttings. Node Culture = promote shoot growth using either intact shoots placed horizontally on the medium or using node segments.

What are node segments? Shoots are then rooted or acclimatized to ex vitro conditions. *Ex vitro – very intuitive. I like it.* Associated with the least genetic variation.

MICROPROPAGATION STAGES

Murashige in 1974 positied three stages for successful micropropagation, though we now recognize five stages.

Stage 1:

- *Donor plant selection*: In vitro responsiveness influenced by phytosanitary and physiological conditions of donor. Best if maintained in lower humidity, drip irrigation and antiobiotic sprays. *In animal culture, they repeatedly tell us to avoid use of antibiotics. I wonder why this is different.* Trimming, cytokinin or gibberellic acid sprays and forcing solutions (2% sucrose and 200mg/L 8-hydroxyquinoline citrate) will all increase explant responsiveness.

Stage 2:

- *Establishment of aseptic cultures*: Use surface disinfested apical meristems, meristem tips or shoot tips from terminal or lateral buds. Screening for internal microbial contaminants is essential. Stabalization in this phase can take 3-24 months. *Wouldn't that be a great way to spend your year. No wonder this is expensive.* The time can be reduced by initial culture in liquid media.

  Factors affecting stage 2 establishment include:

  - *Explantation time*: In reference to period of growth
  - *Explant position on stem*: Terminal bud explants better than lateral

How are they so different that this matters? What else is going on there?
- *Explant size*: Juveniles do better, smaller more dependent on medium supplements (NEED auxin or cytokinin, where usually just enhance – BA most commonly used).
- *Polyphenol oxidation*: Accumulation of these will kill explants. *Clearly that would not be good.*

PGRs for this stage must promote establishment and shoot growth but limit callus formation and adventitious shoot formation.

Stage 3:
- *Proliferation of axillary shoots*: Characterized by repeated enhanced formation of axillary shoots. Cultures are routinely subdivided for further proliferation. Often the most expensive stage. Inverting shoot explants can double or triple axillary shoot production. Selection of cytokinins and the amount for use is done via trial and error it seems, watching multiplication rate, shoot length, and frequency of genetic variation. Auxin may be used, but can be risky because it increases the chance of callus formation. Additionally, when selecting a cytokinin, possibility of adverse carryover to Stage IV should be considered. *After all this work, you wouldn't want to set yourself up for failure!*

Stage 4:
- *Pre-transplant (rooting)*: Preparing for successful transfer to soil. All about elongating the shoots. Expensive. *That seems to be a theme here...*Often try to eliminate this stage due to its high costs. Pre-treating in aqueous auxin prior to transplanting can be beneficial. For woody plants auxin is a must.

Stage 5:
- *Transfer to natural environment*: Ultimate success of the whole procedure rests here. Two reasons underlie most plant failure when transferring from in vitro to greenhouse conditions:
  - Heterotrophic mode of nutrition must rapidly switch to photoautotrophic for survival.
  - Poor water loss control due to epicuticular wax reduction, poorly differentiated mesophyll, abnormal stomata function or poor vascular connection between roots and shoots.

These reasons make good sense. I hadn't thought about the heterotropic to photoautotrophic switch yet, but that would be hard on a poor little transplant. Acclimitizing the transplants through gradual humidity reduction and higher light intensities over 1-4 weeks will promote later survival.

ASEPTIC-MICROPROPAGATION TECHNIQUES

While keeping plant parts clean for disease free culture requires diverse procedures in a commercial operation, a simplified general procedure and culture medium can be used in the home. The few basic supplies are available from local grocery, drug, or health food stores.

The home tissue culture medium described here was tested on African violet leaf/petiole sections, Boston fern rhizome tipes, and wandering Jew shoot tips. The medium was compared to a commercial complex medium and had the same results.

For one pint of medium, mix the ingredients in *Table* together. While constantly stirring, bring the medium to a boil. Continue to boil it until the agar flakes dissolve completely, approximately two to five minutes.

Table. Home Tissue Culture Medium.

Medium Constituent	Amount
Table sugar	1/8 cup
Tap water	1 cup
All-purpose soluble fertilizer	1 cup of stock solution
Inositol tablet (250 mg)	½ tablet
Vitamin tablet with thiamine	¼ tablet
Agar flakes	2 tablespoons

[1] This mixture makes 1 pint of medium.

[2] Prepare the fertilizer stock solution by mixing

¼ Tablespoon of a balanced, water-soluble fertilizer, such as 10-10-10, in a gallon of water.

[3] Inositol is sometimes sold as myo-inositol.

Once the agar flakes dissolve, pour the medium into heat-resistant glass receptacles, such as baby food or canning jars. Pour medium at least 1 inch deep into each jar. Loosely screw on the lids when all of the jars are filled. They are now ready to be sterilized, along with water, tweezers, and razor blades that will be used later. Place the water in heat-resistant jars with lids loosely screwed on or with the foil covers. Wrap the tweezers and razor blades in paper towels.

Sterilize by placing the items in a pressure cooker at 15 psi for 15 minutes. Slowly release the pressure after 15 minutes and place all the items in a clean area. Tighten the lids on the jars with the medium and allow it to cool and solidi*. If the jars won't be used for several days, place a clean layer of foil over them to prevent dust from gathering.

Plant Culture

Micropropagation will only succeed in a sterile environment. Do all transferring of plant material into the jars as quickly as possible in a clean area. Scrub your hands and counter tops with soap and water just before you begin to disinfect plant material. Rubbing alcohol or a 10 per cent bleach solution (one part commercial bleach to nine parts water) can be used to wipe down the working surface. After the medium is cool, prepare the plant material for culture. Plants must be disinfected before placement on the medium because plants usually harbor bacterial and fungal spores. Bacteria and fungi can grow on the medium faster than the plants and dominate the culture. Various plant parts can be cultured, but small, actively growing portions usually result in the most vigourous plantlets. Ferns, for example, are readily propagated by using only one-half inch of a rhizome tip. For other species, one-half to one inch of the shoot tip is sufficient.

If using the shoot tip, remove the attached leaves and discard. Place the tip into a 10 per cent bleach solution for 10 minutes. Be sure that all plant tissue is submerged in the bleach solution. Rinse off excess bleach by placing the shoot tip in sterile water. Once the plant material has been disinfected it should be touched only with sterile utensils. Sterilize utensils with the medium in the pressure cooker or soak them in isopropyl alcohol (rubbing alcohol). Make sure all alcohol has evaporated from the utensils before they touch the plant material. Hasten evaporation, if necessary, using a burning alcohol lamp. Remove any bleach-damaged tissue with a sterile razor blade after rinsing the plant material. Remove the lid of a culture jar and place a plant part in the jar, making sure that it is not completely submerged under the medium. Recap the jar quickly.

Growing On

Place the cultured plant material in a warm, well-lighted location (not in direct sunlight) to encourage growth. Any contamination of the medium will be obvious in three to four days. Remove and wash contaminated culture jars as quickly as possible to prevent the spread of disease to other uncontaminated cultures. Transplant the plantlets into a standard growing medium when they reach a sufficient size. Handle the plantlets carefully because they are moving from a warm, humid environment to a harsher, drier one. After transplanting, water the plants thoroughly and place them in a clear plastic bag for several days. Gradually remove the bag to acclimate the plants to their new environment.

Although this home tissue culture medium and procedure has worked for some plants, success of the process may vary. The precision of the aseptic procedures and types of fertilizer, vitamins, water, and plants may affect the results. If you enjoy intriguing projects, give home cloning a try.

PROPAGATION TO REPAIR DAMAGED WOODY PLANTS

Bridge Grafting

Our landscape plants, especially the fruit trees, get damaged during winter months from rodent activity. Or, because the turfgrass is growing right up to the trunk, damage may be initiated from mower impact. Wrapping the trunk with galvanized wire — "hardware cloth" — will prevent rodent damage, and removing the sod from the base of the tree for at least two feet will eliminate mower damage. Bridge grafting, while not a propagation technique in the strict sense, takes advantage of some of the principles of propagation to save valuable trees. Damage usually becomes evident as the snow melts in the spring. As soon as the "sap flows" or the "bark slips," usually in late April or early May, bridge grafting may be carried out.

Water sprouts or suckers which are about pencil size in diametre can be collected from the tree. Before an area is to be bridge grafted, the ragged edges must first be trimmed away. This is accomplished by cutting into living tissues above and below the point of injury. Scions are then cut so that when in position, both ends will extend into healthy tissue. The scions (at least six will be needed to repair a complete girdling) are held in the place they will occupy over the injury, and a channel for each end is marked on the trunk to facilitate a perfect fit once nailed into place. The scion wood is beveled on both ends and the marked positions on the trunk are cut slightly shorter than the length of the scion. The scions are then inserted into the cuts, keeping the tops oriented to the top of the tree. If a power stapler is handy, use it to secure the bark flaps over the scion ends. If not, wire brads can be hammered in. After inserting, the scions should have a slight bow to them to allow for normal tree movement.

Once all scions are in place, grafting wax or grafting paint is applied over the surface to prevent drying. If buds should sprout on the inserted scions, simply rub them off. The scion inserts will act as conductors of the products of photosynthesis, moving them into the root system for utilization the following spring.

Details of the Bridge Graft. A and B, the Injured Trunk before and after the Scion, Showing Method of Shaping Cuts; D and E, the Complete Graft Prior to Waxing.

Methods of Dividing and Propagating Selected Herbaceous Plants:

- *Alcea rosea*:
  Hollyhock: A biennial that can be grown from seed in summer for blooming the following year. Transplants easily.
- *Alyssum saxatile*:
  Goldentuft: A hardy perennial that can be grown by seed that needs warm temperatures for germination — 70 F — so summer sowing is suggested for bloom the following season. Propagate by division or softwood cuttings in spring.
- *Aquilegia spp*:
  Columbine: A hardy perennial that grows easily from seed. May need light and moist chilling to stimulate germination. Barely cover seeds, and sow when spring temperatures are still cool.

- *Arabis spp*:
  Rockcress: A hardy perennial that is grown from seed sown when soil temperatures are at 68 F needs light to stimulate germination; cover seeds lightly.
- *Aruncus dioicus*:
  Goat's beard: A hardy perennial that is ideal for border or specimen planting. Seed must be cold stratified. Best propagated by division.
- *Asclepias tuberosa*:
  Butterfly weed: A hardy perennial that needs warm temperatures for successful seed germination. Transplanting and division not recommended.
- *Aster spp:*
  Hardy perennials: Seeds germinate at about 68 F. Dig in fall and divide with each section having roots.
- *Astilbe spp:*
  Astilbe: A hardy perennial that is propagated by division in early spring when they are about 1 inch tall. Seed germination is slow and does not come true.
- *Begonia spp*:
  Tuberous begonias: Seed propagation or by tuberous stems that are divided into sections with each one bearing at least one growing point. Leaf, leaf-bud, and short stem cuttings can be easily rooted. Fibrous-rooted begonias — these are the wax and Christmas begonias — can be propagated by leaf or softwood cuttings taken from young shoots in spring or summer. Rhizomatous types, the Rex begonias, are divided into sections via the plants or rhizomes. Propagated by stem or leaf cuttings.
- *Boltonia spp*: A hardy perennial that can be seed sown or divided in spring or fall.
- *Campanula carpatica*:
  Carpathian bellflower: This, and the Canterberry bells, are hardy perennial and biennial plants, respectively. The perennial is propagated by seed, or divided in early spring or August. The Canterberry bells are propagated by seed, under lights, indoors for best results.
- *Canna spp*: These tender perennials are propagated by splitting the rhizome, making sure each division has a generous portion of stem tissue for restarting. Dig and store in fall after frost blackens foliage, then store in a frost-free area during winter. Start indoors in containers in early March for transplanting outdoors when danger of frost is over.
- *Cerastium tomentosum*:
  Snow-in summer: A hardy perennial that can be propagated in summer by softwood cuttings, or by division in the fall.

- *Chrysanthemum spp*:
  (Dendranthemum): A large genus of annuals and perennials. The garden mum produces lateral shoots that develop from the base of the flowering stems, especially if the tops are cut back. When these shoots are 3-4 inches long and are firm but not woody, they can be used as cuttings for rooting under an intermittent mist system. Rooting will be enhanced with the use of a rooting compound like IBA. Otherwise, the plant crowns can be divided in either the spring or fall.
- *Coleus blumi*: Ego builders for those just starting out; root easily from stem cuttings. A frost sensitive annual.
- *Convallaria majalis:*
  Lilly-of-the-valley: Another easily grown, hardy perennial that can be grown from divisions of the rhizome system the plant develops. Dig and divide in early autumn and get planted by late autumn before freeze-up. Individual "pips" or buds on rhizomes can be stored in refrigeration in sphag-num moss and planted in late winter or early spring for blooms in summer.
- *Dahlias*: Plants grow from large tuberous roots, which are dug in the fall before freeze-up and stored over winter in a cool (below 50F) location, covered with vermiculite or sphagnum moss. In spring when new sprouts appear, divide the clumps so that each has at least one sprout. Plant outdoors after frost periods have passed. Dahlias also root from softwood or leaf-bud cuttings.
- *Delphinium spp*: A genus that includes the larkspur, hardy perennials and annuals. Seeds readily germinate; plants can be propagated by softwood cuttings or by root divisions in spring or fall. For better longivity, use seed or cuttings rather than root divisions.
- *Dicentra spp:*
  Bleeding heart: Sow seeds in fall for natural stratification requirements and flowering the following year. Divide clumps in spring or fall; stem cuttings can be rooted if taken after spring flowering. Root cuttings can likewise be taken after flowering.
- *Digitalis spp*:
  Foxglove: Light sensitive seed; do not sow too deep. Propagate in fall by dividing crowns.
- *Gaillardia spp*:
  Indian blanket flower: Another species that has light sensitive seed. Perennial types can be divided in spring or fall.
- *Gypsophila spp*:
  Baby's breath: G. paniculata is a hardy perennial that can be started by seed or by division of crown in spring or fall. Some double-flowering types are grafted.

- *Hemerocallis spp*: Daylily. Divide clumps in spring or fall, separating into rooted sections. Establishes easily and is excellent for soil stabilization.
- *Hosta spp*:
  Plantain-lily: Propagate by clump division in spring.
- *Iris spp*: The rhizome types are divided after bloom, discarding the old portion and keeping only the vigourous side shoots. Cut leaves back to about 4-6 inches. The bulb types are planted in the fall for spring blooms. The original bulb disintegrates, leaving a cluster of different sized bulbs. Separate and grade, using the largest for flowering next spring, with the smaller being used for further development.
- *Lavandula spp*:
  Lavender: Cuttings from side shoots will root in late summer; divide clumps in the fall.
- *Myosotis sylvatica*:
  Forget-me-not: A biennial that is grown as an annual. M. scorpioides is a seed started perennial that can be divided in the spring.
- *Oenothera spp*:
  Evening-primrose: Divide clumps in the fall.
- *Paeonia spp:*
  Peony: Seed propagation is not practical for home gardener, as it takes seven years to produce a flower. Best propagation method is fall division, giving each piece at least three buds or "eyes."
- *Papaver nudicaule*:
  Iceland poppy and other species of perennial poppies: Sow seeds directly in desired location as they do not transplant well. For propagation, cut up roots into pieces in the fall when foliage dies, 3-4 inches long, lay horizontally and cover with about 1 inch of soil.
- *Phlox paniculata*:
  Garden phlox: Although softwood cuttings will root, better results may be realised if root propagated in the fall. Cut all roots off from the crown, leaving just 2 inches remaining. Replant the crown. Cut roots into 2 inches sections and place in flats of sandy soil. Clumps can be divided in spring or fall.
- *Sedum spp*:
  Many species make up this genus: Most are easily propagated by direct sticking of cuttings, or by simply cutting the mat-forming species into desirable sizes and transplanting.
- *Thymus spp*:
  Thyme: Seed is light sensitive; do not sow too deeply. May be increased by division or softwood cuttings taken in summer.
- *Valeriana officinalis*:

Valerian: Hardy perennial. Root is used in sedative preparations. New plants can be obtained from clump division in spring or fall.

ACHIEVEMENTS AND FUTURE PROSPECTS

Tissue culture techniques are being exploited to enhance crop production and to aid crop improvement efforts. Faster clonal multiplication is being exploited on commercial scale for many horticultural species *e.g.* oil palm, mentha, roses, carnation etc. Tissue cultured somatic tissues are now routinely being used for conservation of those species whose seeds are recalcitrant or ones which do not produce seed at all.

Embryo culture has helped in rescuing hybrid embryos enabling the recovery of many interspecific hybrids and haploid plants. Shoot tip (meristem) culture plays a vital which is of great importance in germplasm exchange, and the development of serological techniques for the detection of viruses in plant materials is a great help to the efforts in this direction.

7

Plant Cell Organogenesis

When it comes to plant cell organelles, they are more or less similar to animal cells, except that the latter lacks chloroplast organelles, that are responsible for photosynthesis. Following is a list of organelles found in plant cell.

Nucleus

Nucleus (plural nuclei) is a highly specialized cell organelle, which stores the genetic component (chromosomes) of the particular cell. It serves as the main administrative centre of the cell, by coordinating the metabolic processes like cell growth, cell division and protein synthesis.

Plastids

Plastids are collective terms for organelles that carry pigments. In a plant cell, chloroplasts are the most prominent forms of plastids, that contain the green chlorophyll pigment. Because of these chloroplast plastids, a plant cell has the ability to undergo photosynthesis in the presence of sunlight and synthesize its own food.

Ribosomes

Ribosomes are plant cell organelles that comprise of proteins (40 per cent) and ribonucleic acid or RNA (60 per cent). They are important organelles responsible for the synthesis of proteins. Each ribosome consists of two parts, a larger subunit and a smaller subunit.

Mitochondria

Mitochondria (singular mitochondrion) are spherical to rod-shaped organelles present in the cytoplasm of the plant cell. They break down the complex carbohydrates and sugars into usable forms, for the plant. As mitochondria indirectly supply energy for the plant cell, they are also called as the powerhouse of the cell.

Golgi Body

Golgi body is also referred to as golgi complex or golgi apparatus. It plays a major role in transporting chemical substances in and out of the cell. After the endoplasmic reticulum synthesizes lipids and proteins, golgi body alters and prepares them for exporting outside the cell.

Endoplasmic Reticulum

Endoplasmic reticulum (ER) is the connecting link between the nucleus and cytoplasm of the plant cell. Basically, it is a network of interconnected and convoluted sacs that are located in the cytoplasm. Based on the presence or absence of ribosomes, ER can be of smooth or rough types.

The former type lacks ribosomes, while the latter is covered with ribosomes. Overall, endoplasmic reticulum serves as a manufacturing, storing and transporting structure for glycogen, proteins, steroids and other compounds.

Vacuoles

Vacuoles are the storage organelles that help in regulating turgor pressure of the plant cell. In a plant cell, there can be more than one vacuole. However, the centrally located vacuole is larger than others, which stores all sorts of chemical compounds. Vacuoles also assist in intracellular digestion of complex molecules and excretion of waste products.

Peroxisomes

Peroxisomes are cytoplasmic organelles of the plant cell, which contains certain oxidative enzymes. These enzymes are used for the metabolic breakdown of fatty acids into simple sugar forms. Another important function of peroxisomes is to help chloroplasts in undergoing photorespiration process.

Well! This was a brief information regarding plant cell organelles, their structure and specific functions. As we have seen, coordination of the functions of plant cell organelles is crucial for carrying out the physiological and biochemical functionalities of the plant.

CELL NUCLEUS: STRUCTURE

The structure of a cell nucleus consists of nuclear membrane (nuclear envelope), nucleoplasm, nucleolus and chromosomes. Nucleoplasm, also known as karyoplasm, is the matrix present inside the nucleus. Let's discuss in brief about the several parts of a cell nucleus.

Nuclear Membrane

The nuclear membrane is a double-layered structure that encloses the contents of the nucleus. The outer layer of the nuclear membrane is connected to the endoplasmic reticulum. A fluid-filled space or perinuclear space is present between the two layers of a nuclear membrane. The nucleus communicates with the remaining of the cell or cytoplasm through several openings called nuclear pores. Nuclear pores are the sites for the exchange of large molecules (proteins and RNA) between the nucleus and cytoplasm.

Chromosomes

Chromosomes are present in the form of strings of DNA and histones (protein molecules) called chromatin. Chromatin is further classified into heterochromatin and euchromatin based on the function. The former type is a highly condensed trancriptionally inactive form, mostly present in adjacent to the nuclear membrane. Euchromatin is a delicate, less condensed organization of chromatin, which is found abundantly in a transcribing cell.

Nucleolus

The nucleolus is a dense, spherical-shaped structure present inside the nucleus. Some of the eukaryotic organisms have nucleus that contains up to four nucleoli. The nucleolus plays an indirect role in protein synthesis by producing ribosomes. Ribosomes are cell organelles made up of RNA and proteins; they are transported to the cytoplasm, which are then attached to the endoplasmic reticulum. Ribosomes are the protein-producing structures of a cell. Nucleolus disappears when a cell undergoes division and is reformed after the completion of cell-division.

Cell Nucleus: Functions

Speaking about the functions of a cell nucleus, it controls the hereditary characteristics of an organism and is responsible for the protein synthesis, cell division, growth and differentiation.

Here is a list of the functions carried out by a cell nucleus:

- Storage of hereditary material, the genes in the form of long and thin DNA (deoxyribonucleic acid) strands, referred to as chromatins.
- Storage of proteins and RNA (ribonucleic acid) in the nucleolus.
- Nucleus is a site for transcription in which messenger RNA (MRNA) are produced for the protein synthesis.
- Exchange of hereditary molecules (DNA and RNA) between the nucleus and rest of the cell.
- During the cell division, chromatins are arranged into chromosomes.
- Production of ribosomes (protein factories) in the nucleolus.
- Selective transportation of regulatory factors and energy molecules through nuclear pores.

As the nucleus regulates the integrity of genes and gene expression, it is also referred to as the control centre of a cell. Overall, the cell nucleus stores all the chromosomal DNA of an organism. The environment which we inhabit and live our daily lives, is an incorporation of several kinds of components. These components, although may not share any resemblance, yet are in a mutual dependency with each other.

These components not only refer to the living things, but other elements which are classified as the non-living things on the Earth. Both these elements are important in their own domain in the environment, as they work to maintain the delicate balance in the ecosystem. And this very balance helps to regulate all those things which makes life possible on the Earth. Ergo, with any of these things missing in this system, the biological organization will suffer nothing less than chaotic events. So let's proceed further to understand these types of components and what role do they play in the levels of organization of life.

MERISTEMATIC TISSUES OF ORGANOGENESIS

Meristematic tissues are responsible for the division of new cells... they are zones of actively dividing cells.

Before we proceed, we should first recognize that growth in plants includes two stages:

1. First the production of new cells and,
2. Secondly the expansion of these cells via uptake of water by the vacuole.

Cell division occurs solely in meristematic regions, while expansion may occur anywhere. Thus in a single plant there are zones of young dividing cells, maturing cells, and mature cells. We will go into more detail on meristematic regions when we go over the stem and root individually, but for now an introduction is sufficient

Let us recognize that there are 3 meristematic regions in the plant:

I. Apical meristems are located at the apices or tips - at root and shoot tips and are directly involved in their elongation

They create derivatives which form primary growth:

- The protoderm which forms the outer dermal layer of tissues,
- The ground meristem which forms the cortical cells and
- The procambium which forms the vascular tissue.

In shoots, plants protect their meristems with young leaves and by forming dormant reserve meristems (*i.e.*, buds), that protect their apical meristems with a root cap

Ferns, Gymnosperms and Angiosperms are all terrestrial plants. Tissues therefore arise as a result of a cell or a group of cells dividing, thus giving rise to a large number of cells. In flowering plants these cells undergo changes that change the structure and function of the cell; differentiation has take place. As a result of differentiation, a cell or group of cells can have a specialized function; *i.e.* specialization takes place. Thus, a tissue is a group of cells that have differentiated to perform a specific function.

If a group of cells are more or the less the same, then it is known as a Simple tissue *e.g.* epidermis, parenchyma, collenchyma and sclerenchyma. If, however, the tissue is composed of a number of different cells, then it is known as a Complex tissue such as xylem and phloem.

DIAGRAMMATIC REPRESENTATION OF A PLANT BODY

Ground Tissue is simple non-meristematic tissue (non-dividing tissue) made up 3 cell types:paranenchyma, collenchyma and sclerenchyma.This tissue generally forms either the pith, cortex or bulk of leaf (mesophyll).

Parenchyma Cells

- Mostabundant cells in plants;
- Spherical cells which flatten at point of contact;
- Alive at maturity;pliable, primary cell walls;
- Large vacuoles for storage of starch, fats, and tannins (denature proteins);
- Primary sites of the metabolic functions such as photosynthesis, respiration, and protein synthesis; they are "ready reserves" from which a plant makes specialized cells to meet its changing needs.

SPECIALIZED PARENCHYMA

- *Chlorenchyma*: Photosynthetic cells; have high density of chloroplasts
- *Aerenchyma*: prominent intercellular spaces that improve gas exchange capacity of the tissue; provide maximum support with a minimum metabolic requirement
- *Transfer cells*: specialized for short distance transfer of solutes between cells; have secondary cell walls; they are inner extensions of wall that increase surface area.

Transfer cells occur in areas of high solute transport, such as secretory tissues, which release substances that are produced within the protoplasm and are moved outside, *i.e.*, nectar cells, mucilage in sundews, and resins.

Collenchyma Cells- Colla = Glue, Glisten

- Living protoplasm;
- Unevenly thickened primary cell walls; elongate cells;
- Longer than wide; just beneath epidermis; function to support growing organs, grass, floral parts, and border veins; their nonlignified cell walls can stretch

Sclerenchyma Cells- (skleos = hard)

- Most are dead at maturity;
- Rigid, thick, lignified, nonstretchable secondary cell walls. there are 2 types of sclerenchyma cells:

Sclereids or stone cells we saw in lab- short; variable shape; form hard layers such as the shells of nuts and seed coats; produce the gritty texture of pears

Fibres- long, slender; occur in strands or bundles; tiny cavity or lumen; the different hardnesses of fibres are used to make coarse rope, linen cloth, etc.

DIRECT SHOOT ORGANOGENESIS FROM LEAF EXPLANTS OF CHRYSANTHEMUM

Shoot regeneration from leaf and stem explants of 19 Chrysanthemum cultivars has been studied. Adventitious shoots could be regenerated from leaf and stem explants of 16 cultivars. Most cultivars showed the best shoot regeneration on Murashige and Skoog medium supplemented with 8.9 μM BAP and 2.7 μM NAA (0.5 mg/l). The number of regenerated shoots per explant was more than 10 in some of the cultivars studied. It has not been posssible to regenerate shoots from leaf explants of cultivars 170 and 260. The cultivars 140 and 8010 produced the highest number of shoots. Regenerated shoots can easily be rooted on half strength Murashige and Skoog medium supplemented with 0.98 μM IBA (0.2mg/l). After potting in soil and transfer to the glasshouse normal plants develop. Apparently no somaclonal variation has been observed among the regenerants.

Biolistic particle bombardment is used to transfer desirable genes into difficult to transform plants, but research is lacking on factors that can affect transformation success. This study was conducted to determine if manipulating culture conditions forRhododendron 'PJM Hybrid' and 'Iridon' chrysanthemum leaf explants could alter transient gene expression of β-glucuronidase (GUS) after bombardment. Both rhododendron and chrysanthemum leaf explants were bombarded with 1.0 μm gold particles coated with 2 μg of linear plasmid DNA, containing the GUS gene. Different explant treatments before bombardment were tested, including time in culture on regeneration medium, growing in the presence or absence of 0, 0.09, 0.2, 0.4, or 0.6 M sucrose, and growing in the dark for 0, 3, or 6 days.

Transient GUS expression was histochemically quantified 2 days after bombardment as the percentages of GUS positive explants and the number of blue spots per GUS positive explant. Transient GUS expression was highest when 'PJM Hybrid' explants were incubated on regeneration medium 9 to 12 days before bombardment, whereas transient GUS expression by 'Iridon' explants was relatively high (>42%) for leaves incubated 3 to 18 days before bombardment. Exposure of 'PJM Hybrid' explants to media containing different concentrations of sucrose decreased transient GUS expression by 71% compared to those grown on sucrose-free medium. Alternatively, 27% more 'Iridon' explants transiently expressed GUS compared to those on sucrose-free medium, and the mean number of spots per GUS positive explant increased from 5 to 13.

Transient GUS expression increased almost three fold when 'PJM Hybrid' explants were placed in the dark for six days compared those in the light. In contrast, the dark treatment had inconsistent effects on transient GUS expression by 'Iridon' explants. Manipulating explant culture conditions before biolistic bombardment influenced transient gene expression and illustrated that genotype-specific differences affect transformation potential.

DIRECT AND INDIRECT SHOOT ORGANOGENESIS

Orenia fournieri is also called the wishbone flower, bluewings, or torenia. It is easy to obtain and to grow.

GENERAL CONSIDERATIONS

Growth of Plants

Torenia can be grown from shot cuttings or seed. They should grow for a minimum of 8 weeks after germination. They should start flowering 12 weeks after germination. They are very vulnerable to white fly infestation.

SHOOT ORGANOGENESIS FROM PETUNIA LEAVES

The garden petunia is a popular plant is called a "cultigen" because this hybrid started in domestication. It is a hybrid of 3-4 petunia species. Garden petunias are easy to tissue culture.

GENERAL CONSIDERATIONS

Growth of Plants

Different petunia cultivars and genotypes will have different needs for PGRs, but almost all will produce shoots in vitro. Choose cultivars of the Grandiflora types. Avoid red flowering cultivars and mixed cultivars.

Eight weeks before starting the experiment, plants seeds in a soilless medium. Do not use overhead watering, as this can cause fungal and bacterial contamination. White flies can be a problem as well.

It's interesting that white flies. They were a problem as well. Seedlings should be growing rosettes when explanting. Flowering plants can be used, but may be contaminated. In each petri dish, place two leaf explants adaxial side up (upper surface up).

NON-ZYGOTIC EMBRYOGENESIS

EMBRYOGENESIS

Somatic embryogenesis is the process of a single cell or group of cells initiating the developmental pathway that leads to reproducible regeneration of nonzygotic embryos capable of germinating to form complete plants. Under natural conditions this pathway is not normally followed but from tissue cultures somatic embryogenesis occurs more often and as an alternative to organogenesis for regeneration of whole plants. According to somatic embryogenesis is, initiated either by preembryogenic determined cells or by induced embryogenic determined cells.

In preembryogenic determined cells, the embryogenic pathway is pre determined and the cells appear to only wait for the synthesis of an inducer (or removal of an inhibitor) to resume independent mitotic divisions in order to express their potential. Such cells are found in embryonic tissues (including scutellum of cereals), certain tissues of young in vitro grown plantlets, the nucellus, and the embryo sac (within ovules of mature plants). Induced embryogenic determined cells, on the other hand, require redetermination to the embryogenic state by exposure to specific growth regulators such as 2, 4-D. These cells are dedifferentiated generally in microspore (anther) cultures and callus cultures. Once the embryogenic state has been reached both cell types proliferate in a similar manner as embryogenic determined cells.

Plant lets are then directly produced by following the full embryogenic pathway as a coordinated group of embryogenic determined cells. Some times individual cell or cells from the group may escape and give rise to either embryoids or nodular embryogenic callus (*e.g.* scutellar callus) consisting of proembryoids. These are embryolike structures which are bipolar units and germinate into full plantlets under suitable culture conditions. In dicotyledonous plants totipotent embryogenic cells have been most commonly obtained from explants of embryonic or young seedling tissues.

Excised small tissues from young inflorescences (before maturation of floral primordia) are equally effective for the induction of somatic embryogenesis in cultures. Other explants used are the scutellum, young roots, petioles, immature leaf, and immature hypocotyl. A high yield of embryogenic calli can also be obtained from isolated fully differentiated mesophyll cells or protoplasts in a defined culture medium. Somatic embryos germinate in situ or when they are excised and cultured individually on a fresh semisolid medium.

The essential requirements for the induction and promotion of somatic embryos are primarily established by the use of suspension cultures. The in vitro development of somatic embryos was first observed in carrot suspension cultures was also able to induce somatic embryogenesis in a callus cultured on a semisolid medium. The presence of auxin in the medium is generally essential for embryo initiation. Tissues or calli maintained continuously in an auxin free medium generally do not form embryos; therefore, somatic embryogenesis is achieved in two steps. First, the callus is initiated and multiplied on a medium rich in auxin (2, 4-D, 0.5 mg 1-1) which induces differentiation of localized groups of meristematic cells called embryogenic clumps. Second, the embryogenic clumps develop into mature embryos when transferred to a medium with a very low level of auxin (0.01-0.1 mg,-1) or no auxin at all.

Hence a medium with auxin is called a proliferation medium and one without auxin an embryo development medium. Mature embryos develop in embryo development medium. Mostly 2, 4- D has been used to induce somatic embryogenesis. A substantial amount of nitrogen, usually in reduced form such as ammonium salts, is required for both embryo initiation and maturation. The source of nitrogen can be in the form of complex addenda such as coconut milk, casein hydrolysate, a mixture of amino acids or single amino acid, and ammonium ions.

MS medium contains high levels of nitrogen in the form of ammonium nitrate. Other items, such as potassium, dissolved oxygen, activated charcoal are also found to playa role in somatic embryogenesis. In monocotyledonous plants, most of which are agriculturally and medicinally important, the vegetative parts do not readily proliferate in cultures. Hence explants are best taken from embryogenic or meristematic tissues (young inflorescences and leaves).

Germination of the somatic embryo can occur only when it is mature enough to have functional shoot and root apices capable of meristematic growth. High auxin levels can inhibit development and growth of the shoot meristem and often embryos mature when the embryogenic cell suspension is transferred to a medium lacking auxin. The addition of low level of cytokinin (zeatin 0.1 μm) in combination with ABA may prove beneficial for embryogenesis of low density cell cultures. Various physical factors such as temperature also affect embryo maturation, depending on the requirement of the species. Maturation of somatic embryos proceeds more normally in complete darkness although light seems essential for somatic embryogenesis in some cultures. Somatic embryos germinate on agar medium without growth regulators. Single embryos may profit by the inclusion of low levels of zeatin (0.1 μm) in the medium.

Sometimes loss of morphogenic potential in embryogenic cultures is seen. This may be due to many reasons. Nuclear changes such as polyploidy, aneuploidy and chromosomal mutations in cultured cells may be responsible for the loss of organogenic or embryogenic potential in prolonged cultures. Altered hormonal balance within the cells or tissues may also be associated with decline in the embryogenic potential.

Sometimes the non embryogenic cells of the explant increase under conditions favourable to their growth, resulting in a gradual loss of the embryogenic component during repeated subcultures.

SOMATIC EMBRYOGENESIS FROM SEEDS OF MELON

Scoop out the seeds from a ripe melon and put them into a wire mesh sieve, then with running water over the seeds rub them gently against the mesh, using it to loosen and remove the stringy fibres. Next place the cleaned seeds in a bowl of water, stir it a few times. Some seeds will float to the top....these are immature or sterile melon seeds, they are hollow and/or light-weight and will float to the top of the water. Skim away these bad seeds and discard them. Stir a few more times and repeat the process until no more sterile seeds float to the the top. Drain the water from the remaining seeds. Afterwards, line a heavy plate or baking pan with waxed paper, spread the seeds out in a single layer onto the waxed paper and place it in sunny spot to air-dry.

Stir the seeds occasionally during the next few hours to make sure all sides are exposed to fresh air, this facilitates even drying. After a day in the sun bring the seeds into the house where they continue to dry for another week or two, stir them daily so they dry evenly. If you've got rainy weather the increased humidity can prolong the drying process another week or so. Melons have thick seeds so be sure they are thoroughly dry before packing them for storage.

DESCRIPTION OF ORGANOGENESIS

"The development of adventitious organs or primordia from undifferentiated cell mass in tissue culture by the process of differentiation is called organogenesis. "The formation of roots, shoots or flower buds from the cells in culture in manner similar to adventitious root or shoot formation in cuttings is called organogenesis.

Caulogenesis

Type of organogenesis by which only adventitious shoot bud initiation take place in the callus tissue.

Rhizogenesis

Type of organogenesis by which only adventitious root formation takes place in the callus tissues.

Organoids

In some culture tissues, an error occurs in development programming for organogenesis and an anomalous structure is formed. Such anomalous organs like structures are known as Organoids. Although Organoids contain the dermal, vascular and ground tissues present in plant organs, they differ from true organ in that the Organoids are formed directly from the periphery of the callus tissue and not from organised mersitemoids.

Meristemoids

Meristemoid is localised group of meristematic cells that arise in callus tissue and may give rise to shoots and or roots. They are also termed as nodules or growth centres.

Cytodiffrentiation

In plant tissue culture, during growth and maturation of callus tissue or free cells in suspension culture, few dedifferentiated cells undergo cytoquiescece and cytosenescence and this twin phenomenon are mainly associated with redifferentaition of vascular tissues, particularly tracheary elements. The whole developmental process is termed as cytodifferetiation.

ORGAN CULTURE

It deals with the culture of the isolated organs (roots) under laboratory conditions *(in vitro)*. Different names are given depending upon the organ used for the culture. For instance the culture of roots, endosperm, ovary, and ovule are called as root culture, endosperm culture, ovary culture, etc. It was Skoog (1944), who for the first time suggested that the organogenesis could be chemically controlled. Skoog and Miller (1957) also demonstrated that a high ratio of auxin: cytokinin stimulated the formation of root in tobacco callus, but a low ratio of the same induced shoot information.

EXPLANT CULTURE

The culture of plant parts (explants) is known as explant culture. The explants can be any part of the plant, *e.g.*, the piece of stem, leaf, hypocotyl, etc. The explant cultures are generally used to induce callus or plant regeneration.

Callus Culture

Callus refers to an unorganised mass of cells generally parenchymatous in nature. The unique feature of callus is that the abnormal growth has biological potential to develop normal root, shoots, and embryoids, ultimately forming a plant. Naturally, the callus is formed due to the infection of microorganisms from wounds due to stimulation by endogenous growth hormones, the auxins and cytokinins.

However, it has been possible to artificially develop callus by using tissue culture techniques. Auxins are added to culture medium for callus induction but the nature and quantity of auxin added, depends on the nature and source of explant and its genotype besides other factors. Callus cultures can be maintained for prolonged periods by repeated sub-culturing.

Callus cultures are used for:

- Plant regeneration,
- Preparation of single cell suspensions and protoplasts, and,
- Genetic transformation studies.

FACTORS AFFECTING CALLUS CULTURE

- The source and the genotype of the explant
- Composition of the medium (most commonly used-MS medium)
- Temperature (22-28°C suitable for callus formation)
- Growth regulators, *e.g.*, auxins, cytokinins alone or combination of these.
- Age of the plant- Location of the explant
- Physiology and growth condition of the plant

CELL SUSPENSION CULTURES

Cell suspension is prepared by transferring a fragment of callus to the liquid medium and agitating them aseptically to make the cells free. Single cells can be isolated from either callus or any other part of the plant and cultured in liquid medium using both mechanical and enzymatic methods. Mechanical methods involve grinding of the tissue to a fine suspension in a buffered medium followed by filtration and centrifugation to get rid of cell debris.

The enzymatic method uses the enzymes (pectinase or macerozyme) to dissolve the middle lammela between the cells. After the isolation of the cells, they are cultured by batch cultures or continuous cultures. As the medium is liquid in nature, the pieces of callus remain submerged which creates anaerobic conditions. To overcome this problem, the suspension cultures are agitated by a rotary shaker which disperses the cells and expose them to air.

The advantages of cell suspension cultures over the callus culture:

- The suspension can be pipetted.
- They are less heterogeneous and cell differentiation is less pronounced.
- They can be cultured in volumes upto 1,500 litres.
- They can be subjected to more stringent environmental controls.
- The manipulations for the production of natural products by feeding precursors, is possible.

Batch cultures are initiated as single cells in 100- 250 ml flasks and are propagated by transferring regularly small aliquots of suspension to a fresh medium. Continuous cultures are maintained in a steady state for long period by draining out the used medium and adding fresh medium.

The cell suspension cultures can be used for:

- Induction of somatic embryos/shoots,
- *In vitro* mutagenesis and mutant selection,
- Genetic transformation,
- Production of secondary metabolites.

Mass Cell Culture

Plant cells are refined in specially designed 'plant bioreactors' which essentially do not have a stirrer as plant cells are shear sensitive. In place of stirrer, gas is gently bubbled which provides stirring as well as meet the demand of a higher oxygen supply.

Protoplast Culture

Protoplasts are plant cells without cell wall and can be isolated by using enzymes like cellulases, pectinases) from leaf, seedling, calli, pollen grains, embryo sacs, etc. The protoplasts regenerate cell wall, undergo cell division, and form callus.

The callus can also be subcultured. Some of the examples of plant species that have been regenerated from protoplasts are— Cucumis sativus, Capsicum annum, Ipomoea batata, Glycine max, Chrysanthemum sp.

These cultures are used for:

- Various biochemical and metabolic studies,
- Fusion of two somatic cells to create somatic hybrids,
- Fusion of enucleated and nucleated protoplasts to create Cybrids (cytoplasmic hybrids) and
- Genetic manipulation.
- Drug sensitivity.

Bergmann's Cell Plating Technique

In this technique, free cells are balanced in a liquid medium. Equal volumes of liquid and agar media are mixed and rapidly spread in petri dish, which makes the cells evenly distributed in a thin layer after solidification. After sealing the Petri dishes with parafilm, they are examined under the inverted microscope to mark the single cells. Plates are incubated in dark at 25°C and cell colonies developing from marked single cells, are used to obtain single cell cultures.

EMBRYO CULTURE

Besides, roots, shoots, and pollen, embryos can also be cultured to produce haploid plants. The embryo culture is very useful in conditions where embryo fails to develop due to degeneration of embryonic tissues. It has been used as a routine technique in orchid propagation, in breeding of species showing dormancy.

Embryo Rescue

It has been observed that sometimes, inspite of successful pollination and fertilization, the embryos do not develop. The incompatibility between the embryo and the endosperm or some inherent deficiency also results in the under development of embryo.

These immature embryos can be dissected out from the seeds and can be grown artificially on culture medium. These embryos differentiate into shoot, root and plantlets under culture conditions. This technique of growing immature embryo is termed as 'embryo rescue'. This technique is very useful in hybridisation, breaking dormancy of certain seeds, and to achieve complete growth of embryo into a plant.

Anther and Pollen Culture

Haploid plants possess a single set of chromosomes (gametophytic no of chromosomes, *i.e.*, n) in the sporophyte in contrast to diploids which contain two sets of chromosomes (2n). The existence of haploid plants was reported by Bergner (1921) in Datura stramonium.

Tulecke (1951) cultured the pollen grains of Ginko biloba (gymnosperm) and succeeded in inducing the development of haploid callus. Guha and Maheswari (1964) reported the direct development of haploid embryos and plantlets from microspores of Datura inoxia by the cultures of excised anthers. In 1967, Bourgin and Hitsch obtained the first full haploid plants from Nicotiana tabacum.

Haploid plants are very useful in:

- Direct screening of recessive mutation because in diploid or polyploid screening of recessive mutation is impossible.
- Development of homozygous lines in a short span of time.
- The generation of exclusive male plants by the process of androgenesis using techniques to double the chromosome numbers.
- Production of disease resistant plants by introducing disease resistant genes, *e.g.* Barley accession Q 21681 resistant to stem rust, leaf rust, and powdery mildew.
- Cytogenetic research which includes production of aneuploids, determination of origin and basic number of chromosome numbers, induction of genetic variability.

At present, more than 247 plant species and hybrids belonging to 38 genera and 34 families of dicots and monocots have been regenerated using anther culture technique, *e.g.*, rice, wheat, maize, coconut, rubber trees, etc. The Institute of Crop Breeding and Cultivation (China) has developed the high yielding and blast resistant varieties of rice zhonghua No.8 and zhonghua No. 9 through transfer of desired alien gene.

ANDROGENESIS

In androgenesis, the male gametophyte (the microspore or immature pollen) produces haploid plants by stopping the development of pollen cell into a gamete and forcing it to develop into a haploid plant.

In Vitro Androgenesis

In vitro androgenesis is the formation of sporophyte from the male gametophyte on artificial medium and is most commonly found in family Solanaceae and Poaceae (Graminae).

Direct Androgenesis

In the pollen derived embryogenesis, also called direct androgenesis, the pollen directly acts as a zygote and passes through various embryogenic stages similar to zygotic embryogenesis. Direct androgenesis is very common in many plants of the family Solanaceae and Brassicaceae.

Indirect Androgenesis

The process where the pollen grains instead of normal embryogenesis, divide erratically to develop callus is called indirect androgenesis, *e.g.*, in barley, wheat, Coffee, etc.

The Technique of Anther Culture

The anthers with their filaments are removed from the flower buds after surface sterilisation. Under aseptic conditions, the anthers are excised and crushed in 1 per cent acetocarmine to test the stage of pollen development. The anthers in correct stage of development are separated and inoculated on a nutrient medium. The anther cultures are maintained at 28°C and alternating photo periods of light (12-18 hrs) and darkness (6-12 hrs). The anthers proliferate and produce callus which forms an embryo and the embryo subsequently develops into a haploid plant.

The Technique of Pollen Culture

The pollens are extracted by pressing and squeesing the anthers with a glass rod against the sides of the beaker. The anther tissue debris is removed by filtering the pollen suspension and large and healthy pollen are washed and collected. These pollen are cultured on a solid or liquid medium and the callus or the embryo formed is transferred to a suitable medium to produce a haploid plant.

The factors that affect androgenesis are the genotype and the physiological state of the donar plants, composition of the culture medium, the method adopted in the pretreatment of anthers to develop in to haploid plants and the stage of the microspore or pollen selected for culture

The haploid are identified by looking at their morphological features or by using genetic markers, *e.g.*, a1marker for brown coloured aleurone to detect haploids in maize plant.

Limitations in Haploid Production

- Due to the low frequency of haploid production the selection is very difficult.
- Haploids with deleterious traits frequently develop in cultures.
- It is sometimes difficult to isolate haploids from the culture since the polyploids outgrow haploids.
- The doubling of haploids (diploidisation) does not always lead to the formation of homozygous plant.
- The embryos derived from haploids often get aborted.

ORGANELLES OF EUCARYOTIC CELL

The cells of protozoa, higher plants and animals are highly structured. These cells tend to be larger than the cells of bacteria, and have developed specialised packaging and transport mechanisms that may be necessary to support their larger size.

Use the Interactive animation of plant and animal cells to learn about their respective organelles:

- *Nucleus:* The nucleus is the most obvious organelle in any eukaryotic cell. It is enclosed in a double membrane and communicates with the surrounding cytosol via numerous nuclear pores. Within the nucleus is the DNA responsible for providing the cell with its unique characteristics. The DNA is similar in every cell of the body, but depending on the specific cell type, some genes may be turned on or off - that's why a liver cell is different from a muscle cell, and a muscle cell is different from a fat cell. When a cell is dividing, the nuclear chromatin (DNA and surrounding protein) condenses into chromosomes that are easily seen by microscopy.
- *Nucleolus:* The prominent structure in the nucleus is the nucleolus. The nucleolus produces ribosomes, which move out of the nucleus and take positions on the rough endoplasmic reticulum where they are critical in protein synthesis.
- *Cytosol:* The cytosol is the "soup" within which all the other cell organelles reside and where most of the cellular metabolism occurs. Though mostly water, the cytosol is full of proteins that control cell metabolism including signal transduction pathways, glycolysis, intracellular receptors, and transcription factors.
- *Cytoplasm:* This is a collective term for the cytosol plus the organelles suspended within the cytosol.
- *Centrosome:* The centrosome, or MICROTUBULE ORGANISING

CENTRE (MTOC), is an area in the cell where microtubules are produced. Plant and animal cell centrosomes play similar roles in cell division, and both include collections of microtubules, but the plant cell centrosome is simpler and does not have centrioles.

- During animal cell division, the centrioles replicate (make new copies) and the centrosome divides. The result is two centrosomes, each with its own pair of centrioles. The two centrosomes move to opposite ends of the nucleus, and from each centrosome, microtubules grow into a "spindle" which is responsible for separating replicated chromosomes into the two daughter cells.
- *Centriole (animal cells only):* Each centriole is a ring of nine groups of fused microtubules. There are three microtubules in each group. Microtubules (and centrioles) are part of the cytoskeleton. In the complete animal cell centrosome, the two centrioles are arranged such that one is perpendicular to the other.
- *Golgi:* The Golgi apparatus is a membrane-bound structure with a single membrane. It is actually a stack of membrane-bound vesicles that are important in packaging macromolecules for transport elsewhere in the cell. The stack of larger vesicles is surrounded by numerous smaller vesicles containing those packaged macromolecules. The enzymatic or hormonal contents of lysosomes, peroxisomes and secretory vesicles are packaged in membrane-bound vesicles at the periphery of the Golgi apparatus.
- *Lysosome:* Lysosomes contain hydrolytic enzymes necessary for intracellular digestion. They are common in animal cells, but rare in plant cells. Hydrolytic enzymes of plant cells are more often found in the vacuole.
- *Peroxisome:* Peroxisomes are membrane-bound packets of oxidative enzymes. In plant cells, peroxisomes play a variety of roles including converting fatty acids to sugar and assisting chloroplasts in photorespiration. In animal cells, peroxisomes protect the cell from its own production of toxic hydrogen peroxide. As an example, white blood cells produce hydrogen peroxide to kill bacteria. The oxidative enzymes in peroxisomes break down the hydrogen peroxide into water and oxygen.
- *Secretory Vesicle:* Cell secretions - *e.g.,* hormones, neurotransmitters - are packaged in secretory vesicles at the Golgi apparatus. The secretory vesicles are then transported to the cell surface for release.
- *Cell Membrane:* Every cell is enclosed in a membrane, a double layer of phospholipids (lipid bilayer). The exposed heads of the bilayer are "hydrophilic" (water loving), meaning that they are compatible with water both within the cytosol and outside of the cell. However, the hidden tails of the phosopholipids are "hydrophobic" (water

fearing), so the cell membrane acts as a protective barrier to the uncontrolled flow of water. The membrane is made more complex by the presence of numerous proteins that are crucial to cell activity. These proteins include receptors for odours, tastes and hormones, as well as pores responsible for the controlled entry and exit of ions like sodium (Na^+) potassium (K^+), calcium (Ca^{++}) and chloride (Cl^-).

- *Mitochondria:* Mitochondria provide the energy a cell needs to move, divide, produce secretory products, contract - in short, they are the power centers of the cell. They are about the size of bacteria but may have different shapes depending on the cell type. Mitochondria are membrane-bound organelles, and like the nucleus have a double membrane. The outer membrane is fairly smooth. But the inner membrane is highly convoluted, forming folds (cristae) as seen in the cross-section, above. The cristae greatly increase the inner membrane's surface area. It is on these cristae that food (sugar) is combined with oxygen to produce ATP - the primary energy source for the cell.
- *Vacuole:* A vacuole is a membrane-bound sac that plays roles in intracellular digestion and the release of cellular waste products. In animal cells, vacuoles are generally small. Vacuoles tend to be large in plant cells and play several roles: storing nutrients and waste products, helping increase cell size during growth, and even acting much like lysosomes of animal cells. The plant cell vacuole also regulates turgor pressure in the cell. Water collects in cell vacuoles, pressing outward against the cell wall and producing rigidity in the plant. Without sufficient water, turgor pressure drops and the plant wilts.
- *Cell Wall (plant cells only):* Plant cells have a rigid, protective cell wall made up of polysaccharides. In higher plant cells, that polysaccharide is usually cellulose. The cell wall provides and maintains the shape of these cells and serves as a protective barrier. Fluid collects in the plant cell vacuole and pushes out against the cell wall. This turgor pressure is responsible for the crispness of fresh vegetables.
- *Chloroplast (plant cells only):* Chloroplasts are specialised organelles found in all higher plant cells. These organelles contain the plant cell's chlorophyll responsible for the plant's green colour and the ability to absorb energy from sunlight. This energy is used to convert water plus atmospheric carbon dioxide into metabolisable sugars by the biochemical process of photosynthesis. Chloroplasts have a double outer membrane. Within the stroma are other membrane structures—the thylakoids. Thylakoids appear in stacks called "grana" (singular = granum).

- *Smooth Endoplasmic Reticulum:* Throughout the eukaryotic cell, especially those responsible for the production of hormones and other secretory products, is a vast network of membrane-bound vesicles and tubules called the endoplasmic reticulum, or ER for short. The ER is a continuation of the outer nuclear membrane and its varied functions suggest the complexity of the eukaryotic cell. The smooth endoplasmic reticulum is so named because it appears smooth by electron microscopy. Smooth ER plays different functions depending on the specific cell type including lipid and steroid hormone synthesis, breakdown of lipid-soluble toxins in liver cells, and control of calcium release in muscle cell contraction.
- *Rough Endoplasmic Reticulum:* Rough endoplasmic reticulum appears "pebbled" by electron microscopy due to the presence of numerous ribosomes on its surface. Proteins synthesized on these ribosomes collect in the endoplasmic reticulum for transport throughout the cell.
- *Ribosomes:* Ribosomes are packets of RNA and protein that play a crucial role in both prokaryotic and eukaryotic cells. They are the site of protein synthesis. Each ribosome comprises two parts, a large subunit and a small subunit. Messenger RNA from the cell nucleus is moved systematically along the ribosome where transfer RNA adds individual amino acid molecules to the lengthening protein chain.
- *Cytoskeleton:* As its name implies, the cytoskeleton helps to maintain cell shape. But the primary importance of the cytoskeleton is in cell motility. The internal movement of cell organelles, as well as cell locomotion and muscle fibre contraction could not take place without the cytoskeleton. The cytoskeleton is an organised network of three primary protein filaments:
  - Microtubules
  - Actin filaments (microfilaments)
  - Intermediate fibres.

CELL GROWTH IN INTACT ORGANS

The process of cell growth involves the changes that occur between the meristematic and the mature fully vacuolated state. The meristematic cell is small isodiametric and without a prominent central vacuole. As the cell enlarges a central vacuole develops and the shape changes. Eventually a large parenchymatous cell is established with a thick wall and a thin peripheral layer of cytoplasm. This set of changes is readily studied on the root. It is probable that this is a more satisfactory experimental object than the coleoptile, since with this organ the earlier stages of growth are not readily accessible to investigation. The root, being an organ of unlimited growth, has an apical meristem from which cells are being continuously generated. These cells immediately they are formed commence enlarging, and thus at increasing distances from the apex the tissue is composed of cells in progressively advanced stages of development. A similar situation is involved in the apex of the stem. Here, however, the position is complicated by the development of lateral members from the surface of the meristematic dome. In the root the surfaces of the growing regions are not covered with developing primordia, and this circumstance has been exploited in the development of two sets of techniques for the study of cell growth. Since the surface of the growing region is naked in the root direct observations may be made microscopically on epidermal cells.

The dimensions of epidermal cells may be measured at increasing distances from the apex and the growth of these cells determined over the whole range of development. This technique has been used in principle by Burstrom, Brumfield and by Goodwin. It has been shown that sliding growth is not involved, and changes in surface cells may therefore be taken as representative of corresponding changes occurring in deeper tissues. This general technique has yielded important results, and it has the merit that it involves observations on a more or less uniform tissue. It carries the limitation, however, that only a limited variety of observations can be made with it. It cannot, for instance, yield data regarding metabolic conditions in developing cells. In order to avoid this difficulty another experimental approach has been proposed by Brown and Broadbent. These workers cut successive sections from the apex backwards, made observations on each section, and related the values obtained to the total number of cells in each section. They used a technique for determining the number of cells in a tissue originally developed by Brown and Rickless.

The range of observations that can be made with this technique is clearly greater than that possible with the earlier procedure. It involves, however, two possible sources of error. Cell division does not cease at the same distance from the apex in all regions of the root, and all cells in the section are therefore not exactly at the same stage of development. Secondly, sectioning necessarily involves a destruction of cells the extent of which cannot easily be estimated. The first source of error is not decisive, since the distance over which cell division ceases is small relative to the total distance over which growth occurs, and the average given by the section therefore represents approximately the position for all the cells in the section.

The error from the destruction of cells can be reduced by increasing the length of the section and thus decreasing the proportion of damaged cells. These data are from the observations of Brown and Broadbent. The full development occurs over a distance of about 5.0 mm. from the apex and involves about a twenty- to thirtyfold increase in volume. The enlargement is accompanied by important metabolic changes. In the present connexion it may be noted that the enlargement in the early stages is accompanied by the gradual suppression of the capacity to divide. The zone that stretches from the tip to about 1.5 mm. behind it is that over which mitotic figures are normally distributed. At the same time the observations of Wagner and of Gray and Scholes have shown that the mitotic frequency is greatest at the tip and decreases progressively with increasing distance from it.

Brown has emphasized that the frequency of mitotic figures may not be an index of the relative rate of division. In this case, however, since the frequency decreases to zero it probably is, and the data therefore suggest that after formation at the apex the increase in the average cell size is accompanied by a corresponding increase in the average length of the interphase. As growth proceeds the probability of division decreases, and at a certain point the probability becomes so small as to constitute virtual cessation of the process. The conditions that determine this transition have not been explored, but it is clear that the fact that it occurs indicates that growth in the early stages is accompanied by metabolic changes that tend to suppress division. At about the point at which division ceases a prominent central vacuole appears in the cell. This vacuole, which is probably a development from a dispersed vacuolar system in the meristematic cell, increases in size with the enlargement of the cell.

During growth, however, the percentage water content is increasing, indicating that the volume of the vacuole is increasing more rapidly than the other components, and that the enlargement of the cell is being determined primarily by the absorption of water. At the same time it is evident that the increase in water content is being sustained by relatively smaller increases in other phases of the system. These data indicate that the dry weight of the cell increases about tenfold during the course of growth. Preston and Clark and Wirth have shown that the cellulose content of cells in the coleoptile increases during growth and Wirth has demonstrated corresponding changes with soluble carbohydrates.

The increase in dry weight of the cell in the root is therefore probably due partly to similar changes. It is evident, however, that it is not due only to changes in these constituents. In 1941 Blank and Frey-Wyssling described a series of important observations on cell growth in the coleoptile of maize. They reported an increase in protein content during the enlargement of the cell. Blank and Frey-Wyssling found similar changes occurred during cell enlargement in the hypanthium of *Oenothera*. Kopp calculated that in the root the protein content increased, and Brown and Broadbent demonstrated this condition experimentally in pea roots. The appropriate curve is reproduced from the data of Brown and Broadbent, and it is evident from this that protein increases progressively during growth of the cell, resulting in at least a fivefold increase. At the same time it may be noted that when growth ceases there is a slight but significant decrease in protein, the final level being lower than it is when growth stops, although greater than it is in the meristematic region. The increase in protein necessarily implies an increase in the general metabolic activity of the cell during growth. Kopp calculated and Brown and Broadbent later demonstrated an increase in respiration rate per cell during growth.

The data of Brown and Broadbent suggest, however, that when growth ceases there is a slight fall in respiration which is coincident with the decrease in protein. The increase in protein content might also be expected to increase the quantities of particular enzymes. We have recently been examining the changes in the activities of certain enzymes during the growth of the cell. These data have been obtained by taking successive sections along bean roots, killing the sections by freezing at - 10°C., determining on the killed tissue the activities of certain enzymes, and finally reducing the values for each section to a unit cell basis. The activities of three enzyme systems have been investigated: a dipeptidase which hydrolyses alanylglycine, an invertase, and an acid phosphatase.

The activity of the dipeptidase at increasing distances from the apex of the root of barley has already been examined by Bottelier, Holter and Linderstrom-Lang, and Wanner and Leupold have determined the relative sucrose-splitting capacity in different regions of the root of maize. In the bean roots growing in the conditions of our experiments growth continues over about the first 8 mm. The data show that over this range there is about a fivefold increase in dipeptidase, about a twofold increase in phosphatase, and about a twentyfold increase in invertase activity. Further, it is evident that in about the region where growth in length is ceasing the activities of the respective enzyme systems decrease. Determinations of protein in cells of bean roots are not available, but in view of the earlier observations with peas it is probable that the increase in enzyme activity is due to the differentiation of part of the protein increment into enzyme bodies, and that the decrease in enzyme is similarly partly due to a decrease in protein. It is an important aspect of the situation that the activities of the three enzymes do not change relatively to the same extent.

The relative increases are different and the ratios of the activities of any two of the enzymes therefore change during the course of growth. A similar position is indicated by some observations of Linderstrom-Lang and Holter on the activities of what may be two distinct enzymes cleaving leucyl-glycine and alanylglycine respectively at increasing distances from the tip of the root of barley. The data, although they are not expressed in terms of unit cell number, nevertheless show that the ratio between the two enzymes changes as the distance increases. Clearly since the ratios between the enzyme activities change it is evident that the metabolic pattern must change during the course of growth.

During the growth of the cell a differentiation probably occurs in the metabolic state such that the relative intensities of different reactions change in the course of the process. In particular, these observations on the enzyme complement suggest that the metabolic activities of the young cell are different from those of the mature. This conclusion is consistent with the results of an investigation by Morgan of the amino-acid composition of the proteins in different parts of the growing zone of the bean root. The proteins from sections taken at increasing distances from the apex of the root have been hydrolysed and the hydrolysates examined chromatographically. It is evident that the composition of the proteins varies considerably.

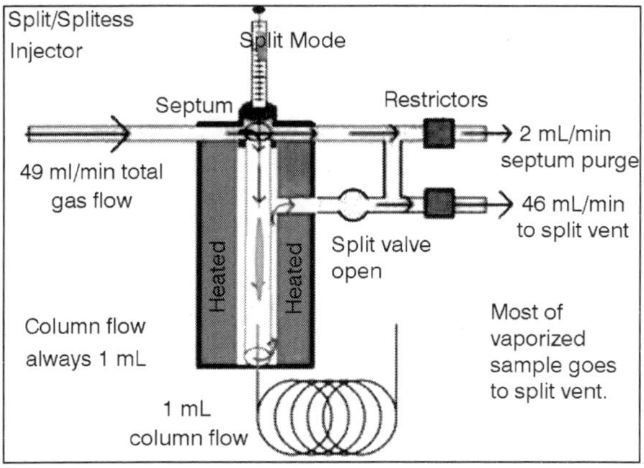

Fig. Chromatographically

Such differences in the composition of the proteins suggest of course corresponding differences in the generalized metabolic pattern. This general conclusion is also consistent with a variety of observations by other workers. Berry and Brock have reported that the respiration of the tip of the onion root is more sensitive to cyanide than is that of the mature tissue, and Kopp has found that per unit protein it is lower at the tip than it is in the mature regions. The data of Brown and Broadbent suggest that the increase in protein per unit increase in volume is greater in the earlier than it is in the later stages of growth. Finally, Burstrom has shown that the reactions of root cells to sugar, temperature and heteroauxin are different in the early and late stages. The growth of plant cells is interesting on account of the fact that the protoplasm is covered with a solid wall. Any irreversible increase of the cell size involves the growth in area of this wall, which consists partly of crystallized substances. The development of cell walls, especially in the cells of higher plants, passes through two different stages.

As long as there is growth in area, they remain thin and flexible. This fine membrane is called the *primary cell wall*, and it is said to grow in area by *intussusception*. As soon as the final shape and size of the cell are reached, the membrane is strengthened by adding new layers to the primary wall.

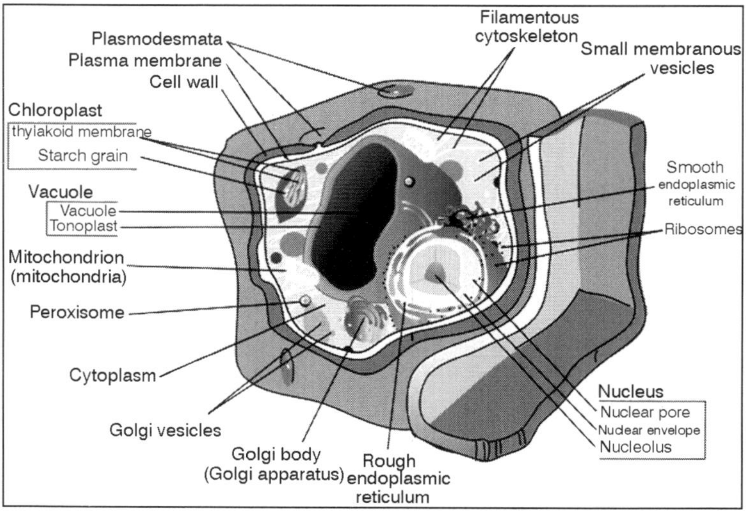

Fig. Plant Cell

During this second stage the cell wall grows in thickness by *apposition*. The combined layers are called the *secondary cell wall*. The thickness of this is often so large that the slender primary wall is overlooked or neglected, especially in materials of technical importance such as wood pulp, textile fibres, etc. Indirect methods of investigation, such as polarized light and X-rays, indicate that the primary wall consists of a loose submicroscopic network of crystallized strands of cellulose or chitin, arranged in a dispersed texture. In the secondary wall the framework is much more compact. This compactness is obtained by a parallel arrangement of the submicroscopic strands forming a highly anisotropic *parallel texture*. Two different types of growth in the area of the primary wall can be distinguished:

TIP GROWTH

Tip growth of filamentous cells such as fungal hyphae, hairs (*e.g.* root hairs), latex tubes, pollen tubes, etc. In this case, only the tip of the cell is involved in growth, and no intussusception occurs away from the end. Sometimes the tip growth of very long cells (*e.g* bast fibres of ramie, which reach a length of over 20 cm.) continue after the formation of the secondary wall has already started in the middle body of the cell.

EXTENSION GROWTH OF CELLS

Extension growth of the cells in elongating tissues. In this case the whole length of the primary wall is said to increase in area by intussusception. There is an enormous bibliography on this subject, because the extension growth is stimulated by auxin, which was considered at the time as the hormone of the extension growth. However, the question how auxins intervene in this process cannot be solved until the mechanism of the growth in area has been cleared up. For a long time it was thought that the extension growth consisted of simple water intake, and the turgor stretch of the cell wall caused thereby.

But it has been shown that this type of cell wall growth is much more complicated, since it involves respiration, biosynthesis, and morphogenesis. Moreover the work needed for the elastic deformation of the growing cell wall is negligible as compared with the total energy available for growth. As extension growth concerns tissues with numerous cells which expand simultaneously at different rates, the geometrical problem how contiguous polyhedra can change their shape without being individualized arises. Krabbe thought that a disintegration of the tissue was inevitable and suggested the hypothesis of *sliding growth*. This implies that the cells of a tissue assume the individuality of independent unities growing along each other like unicellular protophytes. This is an extreme interpretation of sliding growth it is true, but it demonstrates clearly that such an idea is theoretically unsound.

Priestley published other arguments against the possibility of this type of growth; sliding of the two adjacent cell walls would interrupt the channels of pits, so that their orifices would no longer correspond. In consequence one might expect to find a large number of unilateral pits in tissues in which sliding growth had occurred. As this is not the case, Priestley postulated that the adjacent cell faces must grow together at the same rate, and he termed this *symplastic growth*. His theory cannot, however, explain how latex tubes and other specialized cells grow across tissues and thereby come in touch with cells which previously were not their neighbours. This frequently occurs when derivatives of the cambial cells, such as tracheids and fibres, elongate in the tissue.

Two adjacent cells are pushed apart by the tip of a cell which intrudes between them. This has been termed intrusive growth by Sinnot and Bloch or interposition growth by Schoch-Bodmer. The last author emphasizes that only the tips of elongating fibres behave in this way. According to her, the walls of the central body of the cell do not grow any longer, so that the entity of the tissue is secured. Pits are only found in that part of the cell wall, and the fibre tips are free from them.

From this description it follows that different parts of the cell wall must behave differently. A plant cell does not grow as a whole, but in a differentiating cell there are growing and full-grown parts at the same time.

STRUCTURE OF GROWING CELL WALLS

The electron microscope shows that young cell walls have a woven texture of dispersed cellulose microfibrils. The diameter of the microfibrils involved measures about 250 A. However, the electron microscope is said to be unsuited for research in cell physiology and growth because the specimens must be examined completely dried *in vacuo*. Nevertheless, it is possible to take advantage of the results obtained by electron microscopy if they are properly combined with those of other methods. The natural state of the cellulosic framework of the living cell wall can be reconstructed from the electron micrograph if its chemical composition is known. In growing coleoptiles we have found 92.5 per cent water. If this water content is taken as an average for all cell constituents, the cell wall contains only 7.5 per cent dry matter. In corn coleoptiles 32 mm. in length twothirds of this dry material is composed of substances other than cellulose, such as pectins, hemicelluloses, wax. Therefore, the growing cell wall consists of only 2.5 per cent cellulose by weight.

As the density of crystallized cellulose is about 1.55 this is even less by volume. It shows how astonishingly loose is this framework in a growing cell wall, including plenty of space for living cytoplasm and all kinds of incrusting substances. Only when this highly hydrated, spacious network is freed from non-cellulosic substances and completely dried does it furnish the picture of a dense interwoven texture in the electron microscope. The impression of a rather compact framework is further enhanced by the high focal depth of the electron microscope, which yields a sharp picture of microfibrils situated in rather distant planes.

Whereas it is difficult to understand how a woven texture as shown by the electron micrograph can grow by intussusception, where new microfibrils can readily be interlaced. Another handicap to electron microscopy of cell walls has been the method of preparation, which previously consisted in disintegrating cells in a blendor whereby pieces thin enough for electron transmission were obtained. From these specimens it was not possible to locate the pieces examined in the original cell wall, and the micrographs obtained did not show whether the whole wall of a cell or only parts of it are involved in growth. Muhlethaler showed, however, that the purified cell walls of macerated meristems are thin enough for direct examination without cutting them into smaller bits. This is possible because in these membranes he cellulose frame represents only about 2.5 per cent by weight of the living wall, which has a diameter of less than 1¼ in the swollen state.

So the dried cell wall freed of all incrusting substances would be only about 0.025¼ thick, if its mass were spread as a homogeneous film. Yet, according to the porous structure of the wall the dried framework is correspondingly thicker, say 0.075 ¼. This would correspond to a woven texture of about three microfibrils thick. As macerated collapsed cells consist of two superposed walls the thickness of the preparation may amount to 0.15¼, which is an appropriate value for the examination in the electron microscope. The possibility of investigating whole cells proved an opportunity for observing the same spot of the wall alternatively in the electron and in the ordinary microscope. For this purpose the walls are stained with benzoazurin, and the resulting dichroism is examined in polarized light. As the direction of the strong absorption coincides with the main trend of the microfibrils, a mutual check on the observations in the electron microscope and the results of this indirect method is possible.

Thereby an identification of different cell types and the investigation of individual faces of a cell wall are feasible. Our first intention was to elucidate the tip growth in order to observe the formation of new microfibrils at the very tip of the cell where intuesusception growth occurs. *Root hairs* seemed a favourable object for this purpose, because their tip growth is very fast. This research yielded the discovery that the tip of root hairs (*Zea mays*) is covered with a felt-like layer of cellulosic microfibrils. The felt is more than 1 ¼ thick and corresponds to the slime layer which covers the growth region of the hair. The microfibrils are anchored in the interwoven cell wall. The slimy consistency of this layer is due to a large amount of interfibrillar substances with high swelling power, such as pectins. These can easily be demonstrated by pectin dyes (ruthenium red, methylene blue), whilst microchemical reactions, aiming at proving the presence of cellulose, fail.

This is probably due to the fact that the microfibrils found are so well crystallized that no intermicellar spaces are left for the penetration of cellulose dyes. The tip of the root hair is so woolly on account of this extracellular felt that the formation of new microfibrils cannot be observed. It is probable, however, that the cytoplasm which synthesizes the wall substances soaks this system of cellulose microfibrils, which on account of the hydrated state of the pectins must be very loose. As the process of neoformation of microfibrils is hidden in the root hairs, we looked for objects where the growth zone is not covered by slimes, such as hyphae of fungi and pollen tubes. A favourable object is the growing *sporangiophore of Phycomyces* before the tip is inflated for the formation of the sporangium. The chitinous cell walls of fungi have the same microfibrillar structure as those of higher plants; the only difference is that the microfibrils consist of chitin instead of cellulose. Curiously enough both types of microfibrils have approximately the same diameter and both are arranged in loosely woven dispersed or compact parallel textures.

At the tip of the young sporangiophore the texture of the microfibrils is very dense, and is somewhat obscured by some incrusting substance which is difficult to remove. The same difficulty is encountered with the tip of *pollen tubes*. These are very smooth and do not show any structure visible in the electron microscope. Only after extraction with such a strong leaching agent as 10 per cent KOH can the microfibrils be made visible. Therefore, the new wall formed at the tip seems to be cutinized instantly in order to prevent desiccation of the pollen tube which is growing in air. The microfibrils of pollen tubes have a smaller diameter than those of the microfibrils found hitherto in primary cell walls. The supposition that this might be due to the haploid state of the pollen tubes turned out to be wrong. There is no relationship between the size of microfibrils and the number of genoms in the nucleus. This was shown by looking at the haploid alga *Spirogyra,* where very coarse microfibrils occur in the cell wall. Incidentally it was found that the rounded wall of the end cell in a *Spirogyra* thread showed an unexpected tip growth. At the very tip, longitudinal microfibrils spread out from the woven texture of the wall into the open, leaving a gap between them. Along the margin of this gap cross-fibrils are woven into the pattern of the cell wall. It seems that the cytoplasm of the cell tip is so to speak naked while forming the new wall.

CELLS OF MACERATED COLEOPTILES

The extension growth has been studied in the cells of macerated coleoptiles of oat and maize. This classical subject has yielded quite unexpected results. At a very young stage (oat coleoptiles of 5 mm. length) the *parenchyma cells* show an unexpectedly early formation of local secondary bands with parallel texture laid down on the woven primary wall. These reinforcements were found to be located along the edges of the polyhedral cells.

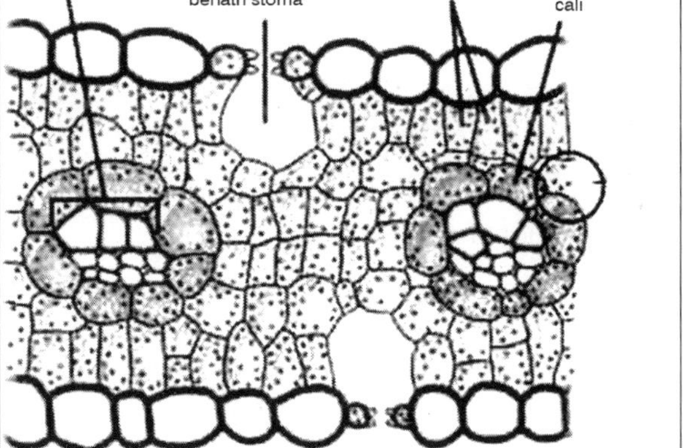

Portion of a Cross Section
Of a Leaf With C_4 Photosynthesis

Fig. Parenchyma Cells

It is difficult to understand how such a parallel-textured strip can extend when the cell elongates. These strips are also visible in the polarizing microscope when stained with benzoazurin. Between the reinforced edges there are thin pitted areas. There again an extension seems quite impossible without distortion of the pits whose elliptic orifices have always a transverse orientation. The elongation of the cells is actually not due to a growth in area of the cell faces already formed, but to a polar or even bipolar tip growth. Since cells are rare in the macerated tissue, this tip growth must go on very rapidly. A gap bordered by longitudinal microfibrils is clearly seen, as well as a margin where transverse microfibrils are woven into this warp. In the living state the tip seems to consist of cytoplasm only, so that the wallbuilding protoplasm protrudes beyond the formed solid wall. This observation excludes any active help of the turgor pressure in cell elongation, because this would force the cytoplasm out of the cell rather than extend the woven framework of the cell wall. Cells with tip growth burst at the very tip when their turgor pressure is increased. This impressive phenomenon of plasmoptysis at the cell tip can easily be observed when growing pollen tubes are immersed in water. *Epidermal cells* prove still more conclusively that cell extension is really tip growth. In oat coleoptiles these cells elongate from about 13 to 2000 ¼ (2 mm.), *i.e.* 150 times within 4 days.

There is a very thick exterior wall which already exists before the cell extension starts and which holds the same thickness when the extension proceeds. The thick outer wall is visible as a broad parallel-textured band at the top of the figure, whilst the rest of the cell is tubular-textured as in the parenchyma cells. At the cell tip this tubular texture grows, and the same manner of growth is used for the elongation of the thick parallel-textured exterior wall. There a much denser tuft of microfibrils is seen which is formed by the living cytoplasm from the cell. Thus the wall lengthens in its definite thickness by bipolar addition of parallelized microfibrils. *Sieve tubes* and *tracheids* also extend by tip growth.

For young fibres and tracheids this was already known. Whilst the cell has already differentiated to a ring tracheid at the bottom of the picture, the tip consists of a veil of interwoven cellulose microfibrils ending at the top in an open tuft of the longitudinal fibrils. It is clearly seen that the cytoplasm forms the warp first, and only afterwards is the weft plaited into the woven primary wall. Finally, the transverse microfibrils exceed those of the longitudinal warp in number. By this type of growth, parenchyma cells, epidermal cells (apart from the outer wall), sieve tubes and the primary wall of tracheids obtain their *tubular texture* with transverse pits and the highest value of anisotropic properties, such as refraction, absorption, strength, etc., in the transverse direction as had been previously established by indirect methods such as double refraction and dichroism.

CELL GROWTH IN ISOLATED FRAGMENTS

The elucidation of the mechanism of cell growth requires, of course, an analysis of the effects of different external conditions on the overall process and of each of its constituent aspects. Attempts have been made to develop such an analysis with intact organs. The interpretation of the results obtained in this experimental situation, however, is a matter of considerable difficulty, since the effects observed may not be due to a direct influence on the growing cells themselves. In the intact root these metabolites may be supplied to the growing cell from mature tissue. Clearly in this situation differences observed on growing cells may not be due to primary effects on these cells, but may represent secondary effects from influences exerted on mature or meristematic cells. This difficulty is avoided by studying the effects of different experimental treatments on isolated fragments composed of growing cells. This technique was first used by Bonner with *Avena* coleoptile tissue. This worker excised a 3 mm. fragment from the body of a coleoptile which had not reached the mature state, and subsequently observed the increase in length of the fragment in different media. A similar technique has since been used by Christiansen and Thimann with fragments from pea internodes and by Brown and Sutcliffe with fragments excised from the extending zone of the root. This technique is an important one and has given valuable results. Nevertheless the limitations of it as it has hitherto been used require some consideration.

It is an important feature of the technique that the fragment is normally excised from an organ that has completed most of its development. Thus, although the mature coleoptile may be about 4 cm. long, the fragment is frequently taken from an organ which is about 2 cm. long, and the cells of which would therefore normally only grow to about double their size. In the overall growth of the coleoptile the increase in cell size must be considerably greater than this. Avery and Burkholder have shown with the *Avena* coleoptile from 4.0 to about 10 mm. growth is accompanied by cell division, and that from 1.O to 4.0 cm. growth is entirely by cell enlargement. Thus after cell division ceases there is at least a fourfold increase in length, and in view of observations on the root it is possible that cell size is increasing in this phase also, giving an overall increase in size greater than fourfold. A similar situation is involved with the pea stem and root fragments. The root segments of Brown and Sutcliffe are taken from the zone which stretches from 1.5 to 3.0 mm. from the tip. The whole growth zone only occupies about 5.0 mm., and therefore at the time of excision at least the more mature cells in the fragment have completed most of their growth. The maturity of the tissue from which the fragments are taken no doubt accounts at least partly for the limited growth normally given under experimental conditions.

In favourable circumstances fragments from the coleoptile may double in length and those from the root quadruple. These increments, however, are considerably smaller than the normal twentyfold increase in volume with a tenfold increase in length that occurs when the cells are growing in an intact organ. The limited increases observed are of course highly significant with respect to the final stages. On the other hand, in view of the results of observations on the changes in the enzyme complement during growth it is doubtful if results obtained with the final stages are applicable to the earlier. Since the metabolic pattern changes as the cell enlarges it is of course probable that the conditions that limit growth are different in the different phases of the process. This has been demonstrated by measuring the growth during a

24 hr. period of successive millimetre fragments excised from the root of the bean. The fragments have been cultured with a 2 per cent sucrose medium on sintered glass disks. It increases with increasing distance from the tip until about 3.0 mm. and then decreases.

Clearly while it is the sugar supply that limits growth in the 3-4 mm. fragment, it is not this factor which controls the situation in more immature tissue. The growth after 4.0 mm. is less intense, since the cells are approaching the mature state. Thus the evidence suggests that observations with the excised fragments as they have been used hitherto are relevant only to an analysis of the final stages of growth, and it is within this limitation. Recent work has emphasized the importance of the metabolic state of the tissue in the determination of the course of cell growth.

Frequently, however, the rate and total amount of growth are considered as being linked to certain metabolic processes in the sense that an acceleration of the metabolic process leads to a corresponding acceleration in growth, and, conversely, that an inhibition of the metabolic process depresses growth. Evidence is assembled here, however, which suggests that growth may be interpreted in terms of two complementary sets of processes. One set when accelerated tends to promote growth, and a second when accelerated tends to depress it. Whatever the metabolic processes involved are, however, it is clear that they can only affect the growth process through a comparatively limited set of physiological conditions, and stimulations or inhibitions of growth must be interpreted immediately in terms of these conditions. The increase in volume of the fragment is determined of course primarily by the absorption of water which as in any other vacuolated cell is the resultant of the conditions tending to draw water into the cell and of the restraining pressure exerted by the wall.

Water may flow into the cell either along an osmotic gradient or as a result of a possible secretion into the vacuole which involves the consumption of metabolic energy. The importance of the osmotic gradient has been emphasized by Thimann and Schneider, who have shown that when coleoptile fragments are immersed in a solution of mannitol which is not absorbed growth is depressed. In view of this it is evident that if growth is to proceed at a high rate the situation requires the continued absorption and accumulation of solutes in the vacuole, and the rate of solute absorption particularly when fragments are immersed in fluids of high osmotic pressure is clearly an important aspect of the situation.

The restraining pressure of the wall in the earlier phases of the investigation of cell growth was defined in terms of elasticity and plasticity. These are no doubt important aspects of the situation. It is evident, however, that the expansion of the wall and therefore the restraining pressure is determined primarily by the synthesis and deposition of wall constituents. Bonner showed with coleoptile fragments immersed in water or in fructose solutions that during growth there is a synthesis of wall material, and Brown and Sutcliffe found with root fragments that extensive growth only occurs when cellulose is being synthesized. At the same time it is probable that the synthesis of cellulose is only effective in promoting expansion when the physical state is such that the deposition of the cellulose in the wall can occur. Gorter has recently shown that in root hairs which have been treated with triiodobenzoic acid conditions are induced which do not affect the synthesis of cellulose, but prevent the intussusception of it in the primary wall.

Thus the growth of the fragment immediately depends on the operation of at least three and possibly four physiological conditions:

- A possible secretion mechanism for water,
- The absorption and accumulation of solutes,
- The physical state of the wall
- The synthesis of wall material.

These conditions are controlled by metabolic states, and at least some metabolic processes when accelerated modify the physiological conditions in such a way as to promote growth. Several workers have shown that a vigourous respiration is necessary for growth to occur. Bonner found with coleoptile fragments that extension did not occur in the absence of oxygen or in the presence of cyanide.

Brown and Sutcliffe have shown with root fragments that growth is limited when the partial pressure of oxygen is reduced below that of air. The promoting effect of a vigourous respiration is, however, a complex phenomenon, and in view of results considered below it is probable that it involves two aspects. It is probable that it is a necessary condition for the promotion of certain other processes on which growth immediately depends, and that certain of the phases of respiration are directly linked to growth. That these two aspects are involved is suggested by a consideration of the effects of sugar and potassium on growth.

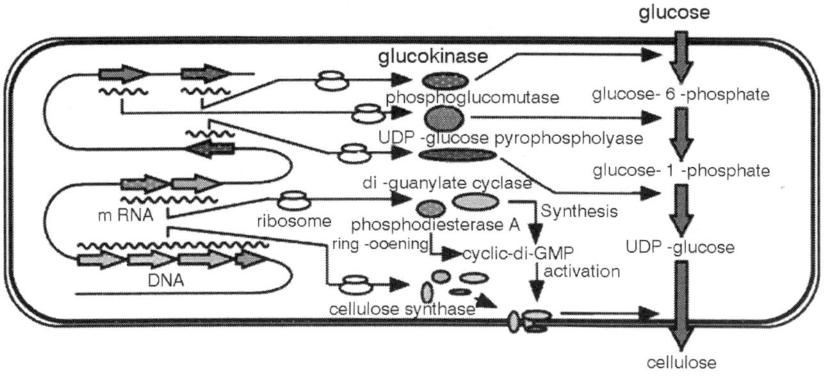

Fig. Cellulose Synthesis

Thimann and Schneider reported that the growth of coleoptile fragments is enhanced by the provision of sugar, and Bentley has recently made a similar observation. Brown and Sutcliffe reported that in the absence of sugar the growth of root fragments is limited and that vigourous growth is only given in the presence of this nutrient. These workers analysed the sugar effect in terms of absorption and accumulation of sugar, synthesis of cellulose and respiration. They found that all these processes tend to increase with increasing concentration of sugar, but only when the cultures are exposed to air.

In low partial pressures of oxygen they found a low respiration, a low absorption, no accumulation, and no cellulose synthesis. It was suggested that a high partial pressure of oxygen promotes a vigourous respiration which stimulates absorption. The rapid absorption in turn promotes the accumulation of sugar which establishes the osmotic gradient along which the absorption of water can occur. The rapid sugar absorption also provides the conditions for the synthesis of cellulose and thus stimulates the expansion of the wall. This interpretation is consistent with the observation that there is no quantitative relation between respiration and growth. In this context respiration provides the necessary condition for the promotion of sugar accumulation and cellulose synthesis and is not directly linked to them. At the same time these are energy consuming processes, and a specific mechanism is presumably involved in effecting the energy transfer.

Bonner finds with coleoptile fragments that sodium arsenate inhibits growth but has little effect on respiration, and he suggests that this is due to the inhibition of the transfer of energy-rich phosphate bonds generated in respiration to reactions immediately involved in growth. In the present connexion it may be suggested that respiration provides a source of energy-rich compounds which become involved in reactions immediately connected with sugar accumulation and cellulose synthesis. The effect of potassium suggests that there may also be a direct link between respiration and growth.

The influence of potassium in the growth of coleoptile tissue is uncertain; Thimann and Schneider found a stimulation, Frey-Wyssling an inhibition, and Bentley an inhibition. Whatever the position may be with coleoptile tissue, with root tissue it is clear and unequivocal. Brown and Sutcliffe found that in the presence of potassium growth and respiration are enhanced while cellulose synthesis and the absorption of sugar are not.

The enhanced growth cannot be attributed to a greater synthesis of cellulose. Also since the absorption of sugar is not increased when growth and consequent water absorption are raised, it follows that the enhanced growth occurs with a lower internal concentration of sugar. Brown and Sutcliffe suggest this might be due either to the adjustment of the internal osmotic pressure through an accumulation of potassium or to the stimulation of a secretion mechanism such as that proposed by Bennet-Clark, Greenwood and Barker. It is significant, however, that potassium has no effect either on growth or respiration in the absence of sugar. Cartwright has recently examined the internal potassium concentration of the fragments used by Brown and Sutcliffe in the presence and absence of sugar and finds there is little difference. This suggests that the internal potassium concentration as such can have little effect on the osmotic gradient, and that the accelerating effect of potassium on growth is due to the stimulation of a water-secreting mechanism acting through a primary stimulation of some phase of the respiratory cycle. The interpretation proposed here would involve an intimate connexion between some limited phase of respiration and the secretion of water in the sense that an inhibition of this phase alone would arrest the secretion of water. Commoner and Thimann have provided evidence for an intimate link of this general type between growth in the coleoptile and respiration.

They find that 10-5M-iodoacetate arrests growth but only depresses respiration by about 10 per cent, and they suggest that growth is controlled by this fraction of the total respiration to which it is linked. A vigourous respiration may also be involved in another connexion. During growth in the intact organ there is a net increase in protein content. Even if the protein content does not increase, however, it is probable that protein synthesis must occur, and that this synthesis is necessary for the promotion of growth. The synthesis is an energy-requiring process, and can only occur when respiration is vigourous. The growth may be enhanced by the stimulation of certain metabolic processes. It is clear, however, from curves, that the overall growth process cannot be analysed in terms of such processes alone.

The effects of different concentrations of the same nutrient on the final lengths of the fragments suggest that growth is also determined by a complementary set of processes the stimulation of which tends to arrest it. The final length in 0.25 per cent sucrose is lower than it is in 2 per cent. The lengths do not converge to the same limiting value at different rates depending on the external concentration. A similar situation is shown by the data of Schneider on the effect of different concentrations of sugar on the growth of coleoptile fragments.

If the stimulation of growth depended on the stimulation of a particular set of metabolic processes, with isolated fragments growth might be expected to continue until one or a complex of conditions became limiting. The limiting condition might be such as the protein level, the distance of the nucleus from the two ends of the cell, or some property involved in the geometry of the system. Such a limiting condition, however, would be common to all members of an experimental series, and with different levels of a single nutrient they might all be expected to reach the same final length, but at rates which are directly related to the external nutrient concentration. Clearly some other factor is involved in the process, and this it may be suggested is the operation of an arresting mechanism which stops growth before a hypothetical maximum length can be reached in at least some series. Other evidence is available which supports this conclusion. It has been shown that in the coleoptile the same concentration of heteroauxin is less effective if added 12 hr. after excision than it is if applied immediately. Similarly with root fragments, if these are transferred to water for 6 hr. and then to a sugar solution, the final length is less than it is if the sugar is supplied immediately. Clearly in the interval during which the stimulant has not been available changes have occurred which tend to restrict subsequent growth. Secondly, there is evidence that with certain low partial pressures of oxygen the capacity for growth is retained longer than it is in air. When the oxygen level is very low the conditions of anaerobiosis have the normal effect of damaging the tissue. In 10 per cent oxygen, however, the arresting mechanism tends to be inhibited. In a series of experiments conducted by Sutcliffe fragments of maize roots were cultured in air and in an atmosphere containing 10 per cent oxygen, and at intervals during the experimental period cultures were transferred from the lower to the higher partial pressure.

With regard to continuous exposure to air and to an atmosphere containing 10 per cent oxygen, it is evident that while the rate of growth is lower in the lower partial pressure the time during which growth proceeds is very much greater. Clearly in the lower partial pressure the processes which promote growth are depressed giving a lower rate, while those tending to arrest growth are inhibited giving a longer period of growth.

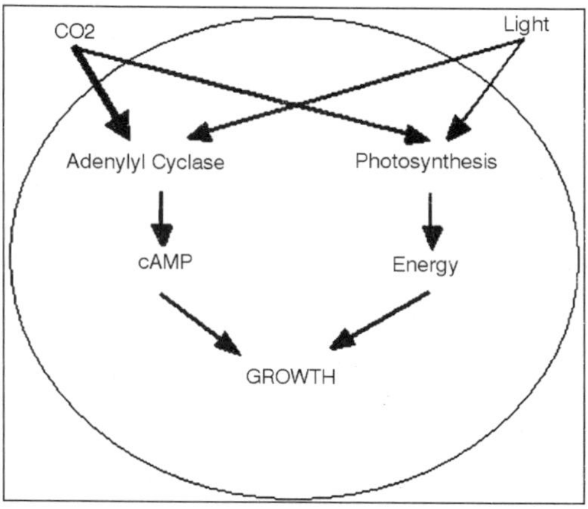

Fig. Various Metabolic Processes

The effects of transferring from the lower to the higher partial pressure are consistent with this interpretation. After transference the rate is accelerated, but this occurs after a period during which, although some growth has occurred, the growth-arresting processes have been inhibited. The net result is that the final lengths of the fragments are considerably greater than they would have been if they had been continuously in air. Thirdly, it has been shown with root fragments that before growth ceases there may be a decrease in respiration. In intact roots it is probable that there are decreases in protein, the activities of certain enzymes, and respiration, either before or at about the time when growth ceases.

The evidence suggests that progressive changes continue during growth, which when they have reached a certain critical level depress the rates of various metabolic processes, and it is as a result of such decreases that growth ceases. The data obtained with isolated fragments which suggest the operation of an arresting mechanism do not indicate the nature of that mechanism. It may be suggested tentatively, however, that this involves the progressive inactivation of an enzyme system, in which case growth may be interpreted in terms of the rate per unit quantity of an enzyme or enzyme complex and the simultaneous absolute or relative reduction in the total quantity of this complex during growth.

In this situation it may be taken that the rate of the reaction per unit quantity of enzyme remains constant. The curve represents the change in the content of enzyme with time. Clearly the distance from this curve to the abscissa represents the rate of growth, and the area under the curve the total amount of growth made. It is immediately evident that at any one time the growth rate is higher, the total amount of growth made is greater and growth proceeds for a longer time. The lengths of the vertical lines here represent the rate per unit quantity of enzyme. It is evident that when the total amount and the rate of growth are affected by the rate per unit quantity of enzyme and by the rate of reduction of the total quantity, then the only factor that can affect the time of growth is the rate of reduction of the quantity of enzyme, and if the time of growth is not affected then no effect on the rate of inactivation can be involved.

Thus it may be suggested that since potassium increases rate and final length but has no effect on time this is due to a stimulation of the promoting mechanism. On the other hand, sugar which undoubtedly stimulates the promoting mechanism, since it increases the time of growth, probably also inhibits the arresting mechanism. It may be emphasized, however, that the present interpretation suggests that an enhanced growth in terms of rate and time could occur by a simple inhibition of the arresting mechanism only. It has been assumed that the rate of inactivation increases with time. It is evident, however, that the general character of the interpretation is not affected by the form of the inactivation curve. The arresting mechanism has been considered above in terms of a reduction in a total quantity of enzyme involved in growth.

0This is suggested by the changes in certain metabolic processes (notably respiration) that occur during the course of growth. Further, measurements have been made of the changes in the activities of certain enzymes, and it has been found with root fragments in certain instances that there may be a decrease in invertase activity and an increase in acid phosphatase activity.

Thus changes during growth do occur in the activities of particular enzymes. These changes may be due to the production of specific inhibitors or to a transformation of one enzyme into another, in either case leading to a reduction in the total quantity of active enzyme. Another possibility may, however, also be considered. It is that growth arrest may be due to a relative and not an absolute decrease of a particular enzyme complex.

If two enzymes increase that compete for the same substrate the reaction catalysed by one enzyme might be virtually suppressed by a very much greater relative increase in the second. It is possible that the inactivation is of this type in the intact root, where there is a considerable increase in total protein during growth. On the other hand, an absolute decrease may be involved in the arrest of the growth of the fragments. It was suggested above that the growth of fragments must be determined by four primary physiological conditions which in turn are regulated by metabolic states. It is clear, however, that the states that arrest growth must operate through these conditions no less than do those which promote it. The rate of growth of the root fragments used by Brown and Sutcliffe decreases with time. This decrease, is an expression of the increasing effect of the arresting mechanism, and should therefore be reflected in a corresponding change in one or more of the primary physiological conditions.

Brown and Sutcliffe, however, found that during growth of the fragments there are no changes in the rates of sugar absorption, sugar accumulation and cellulose synthesis. It is probable, therefore, that the effect of the arresting mechanism is expressed either through a change in the physical state of the wall or in the water-secretion mechanism, the effect of which might of course be determined by a change in the permeability of the protoplast to water.

CELL-WALL GROWTH IN DIVIDING CELLS

How does the cell wall behave when the cell divides? In very young coleoptiles there is still some mitotic activity. We could therefore photograph a certain number of dividing cells. The structure of the expanding cell plate is difficult to observe, because it is inside the enveloping membrane of the mother cell. But the constriction of this wall in the middle of the cell is clearly visible. Its microfibrils are partly bent inwards, where they meet the cell plate. Thereupon they are interwoven with the microfibrils of the corresponding lamella of the double-laminated cell plate.

A certain number of the microfibrils of the original longitudinal wall are left over and decorate the margin of the two daughter cells with their submicroscopic fringes. These microfibrils are gradually dissolved and disappear. This proves that existing cellulose fibrils can be hydrolysed near by the place where other microfibrils are synthesized. Whether these two antagonistic processes are catalysed by the same enzymatic system is open to discussion. The effort of our laboratory to elucidate the mechanism of extension growth has shown that, in the object investigated, there is no difference between tip growth and 'cell extension'. This had already been postulated previously. But it was thought then that the primary wall could expand by loosening its existing framework and subsequently inserting new microfibrils. In order to obtain a local loosening, certain microfibrils had to be dissolved by hydrolysis.

That a local 'melting down' of microfibrils is possible has been proved in this study, when we observed how the ring of microfibrils left over as fringes along the border of the new wall in divided cells gradually disappears. Thus, plasticizing of an existing cellulose frame by dissolution of microfibrils is possible. But in the observed cases of cell extension it does obviously not occur. Growth in area does not proceed by intussusception but by a terminal increment, just as the area of a cloth is enlarged by weaving its margin. In so doing the cytoplasm must somehow protrude beyond its solid case.

As the coleoptiles investigated represent the classical subject for the study of extension growth, there is some danger of generalizing these findings to all cases of growth in area of cell walls and of banishing the concept of intussusception. Although most elongating cells probably grow in the way here described, there are still some cases which must be thoroughly investigated before a general theory of cell-wall growth can be formulated. It will be very interesting to find out how tip growth proceeds in the very fast extension growth of grass filaments.

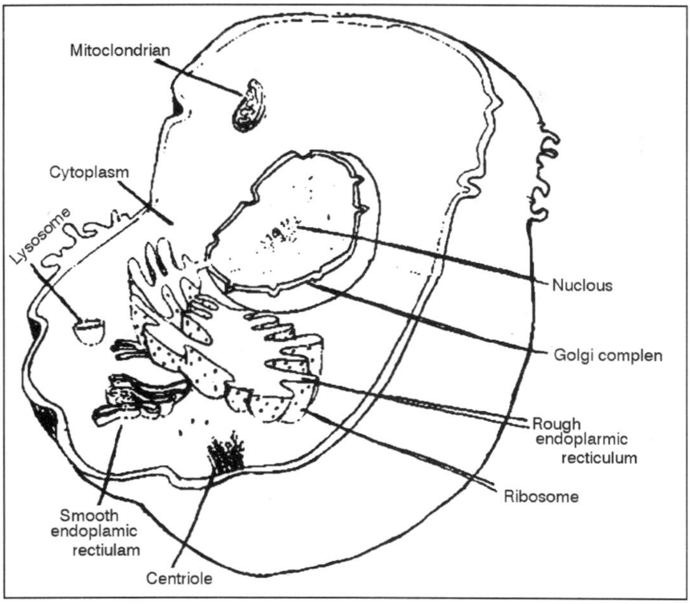

Fig. Cytoplasm

Further, there is an enormous growth in area of the wall when the descendant of a narrow cambial cell differentiates to form a large-bored vessel member. Here no tip growth is possible, but the cell surface must increase along some longitudinal line or lines, where plasticizing of the existing wall is necessary. The same is true when a cell produces some local outgrowth (*e.g.* a root hair). At the place where this reactivated tip growth starts, the existing framework must have been loosened somehow beforehand. Since the texture of primary walls is so spacious, there are many possibilities of rendering the wall softer by increasing the interfibrillar ('incrusting') substances capable of swelling, or by intercalating new microfibrils.

This would be possible where living cytoplasm is located within the growing cell wall. Our study seems to indicate, however, that the cell wall is only soaked with living matter locally. Thus intussusception is restricted to small areas, and the cytoplasm retires very soon from the woven framework. All these considerations show that the growth of the plant cell wall is an autonomic process. Exterior forces such as turgor pressure cannot play an essential role; they are only involved in so far as they influence the activity of the cytoplasm. The extension growth of the cells in coleoptiles is a *bipolar tip growth.* The growing cell wall is living: it must be soaked with cytoplasm which synthesizes microfibrils *in situ*. In dividing cells it bends them to meet the cell plate on which they are fastened. Therefore the growth of the cell wall is an autonomic process which is governed by the morphogenetic faculties of the living protoplast.

8

Cloning of Plants

Plant cloning is the production of a cell, cell component, or plant that is genetically identical to the unit or individual from which it was derived. The term "clone" is derived from the Greek word klon, meaning a slip or twig. Hence, it is an appropriate choice. Plants have been "cloned" from stem cuttings or whole-plant divisions for many centuries, perhaps dating back as far as the beginnings of agriculture.

HISTORICAL BACKGROUND

In 1838 German scientists Matthias Schleiden and Theodor Schwann presented their cell theory, which states, in part, that all life is composed of cells and that all cells arise from pre-existing cells. This theory formed the basis for the concept of totipotency, which states that since cells must contain all of the genetic information necessary to create an entire, multicellular organism, all of the cells of a multicellular organism retain the potential to recreate, or regenerate, the entire organism. Thus was the basis for plant cell culture research. The first attempt at culturing isolated plant tissues was by Austrian botanist Gottlieb Haberlandt at the beginning of the twentieth century, but it was unsuccessful. In 1939 Professor R. J. Gautheret and colleagues demonstrated the first successful culture of isolated plant tissues as a continuously dividing callus tissue.

The term callus is defined as an unorganised mass of dividing cells, such as in a wound response. It was not until 1954, however, that the first whole plant was regenerated, or cloned, from a single adult plant cell by W.H. Muir *et al.* Thereafter, an increased understanding of plant physiology, especially the role of plant hormones in plant growth and development, contributed to rapid advances inplant cell and tissue culture technologies in the 1970's and 1980's. Many plant species have been successfully cloned from single cells, thus demonstrating and affirming the concept of totipotency.

Horticulture

By far, the greatest impact of cloning plants *in vitro* (Latin for "in glass," meaning in the laboratory or outside the plant) has been on the horticultural industry. In the 1980's plant tissue culture technologies propagated and produced many millions of plants. Today, many economically important plants are commonly propagated via tissue culture techniques, including vegetable crops (such as the potato), fruit crops (strawberries and dates), floriculture species (orchids, lilies, roses, Boston ferns), and even woody species (pines and grapes).

The advantages of plant cell, tissue, and organ culture technologies include a more rapid production of plants, taking weeks instead of months or years. Much less space is required (square feet instead of field plots). Plants can be produced year-round, and economic, political, and environmental considerations that hamper the propagation of regional or endangered plant species can be reduced. The disadvantages include the high start-up costs for facilities, the skilled labour required, and the need to maintain sterile conditions. Two other significant considerations must be considered as a result of plant propagation technologies. As illustrated by the Irish Potato Famine of the 1840's, the cultivation of whole fields of genetically identical plants (monoculture) leaves the entire crop vulnerable to pest and disease infestations.

The second important consideration when generating entire populations of clones, especially using tissue culture technologies, is the potential for introducing genetic abnormalities, which then are present in the entire population of plants produced, a process termed somaclonal variation.

Biotechnology

An absolute requirement for genetic engineering of plants is the ability to regenerate an entire plant from a single, genetically transformed cell, thus emphasizing the second major impact of plant cell culture technologies. In 1994 the U.S. Food and Drug Administration (FDA) approved the first genetically modified whole food crop, Calgene's Flavr Savr tomato. This plant was produced using what is termed anti-sense technology.

One of the tomato's genes involved in fruit ripening was reversed, thus inactivating it and allowing tomatoes produced from it to have significantly delayed ripening. Although no longer commercially marketed, the Flavr Savr demonstrated the impact of genetic engineering in moving modern agriculture from the Green Revolution into what has been termed the Gene Revolution.

Other examples of agricultural engineering exist today, such as Roundup Ready Soybeans, engineered to resist the herbicide used on weeds where soybeans are grown, and BT Corn, which contains a bacterial gene conveying increased pest resistance. Since 1987, the U.S. Department of Agriculture (USDA) has required field testing of genetically modified crops to demonstrate that their use will not be disruptive to the natural ecosystem. To date, thousands of field trials have been completed or are in progress for genetically modified versions of several crop species, including potatoes, cotton, alfalfa, canola, and cucumbers.

CUTTINGS OF CLONE PLANT

The simplest way to clone a plant involves taking a cutting. A branch from the parent plant is cut off, its lower leaves are removed, and the stem is planted in damp compost. Plant *hormones* are often used to encourage new roots to develop.

The cutting is usually covered in a clear plastic bag at this stage to keep it moist and warm. After a few weeks, new roots develop and a new plant is produced. The method is easy enough for most gardeners to do successfully.

Tissue Culture

Another way of cloning plants is by tissue culture, which works not with cuttings but with tiny pieces from the parent plant. Sterile *agar jelly* with plant hormones and lots of nutrients is needed. This makes tissue culture more expensive and difficult to do than taking cuttings.

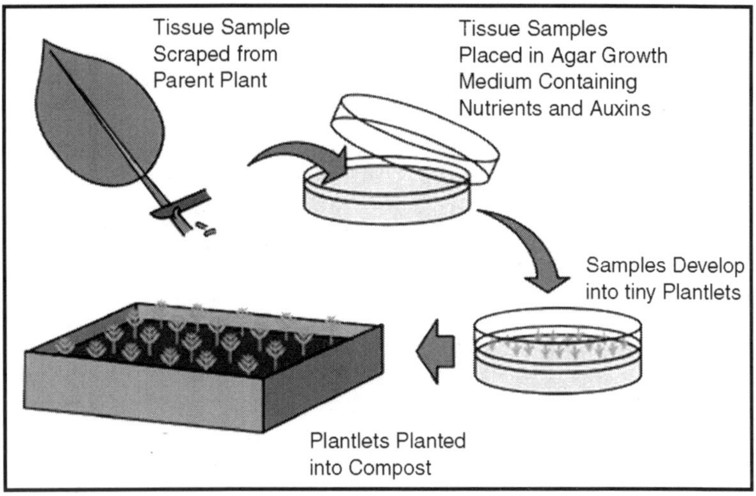

Fig. Cloning Plants by Tissue Culture.

Tissue culture involves the following steps:

- Small amounts of parent tissue or a number of cells are taken
- The plant material is transferred to plates containing sterile nutrient agar jelly
- Plant hormones are added to stimulate the cells to divide
- Cells grow rapidly into small masses of tissue
- More growth hormones are added to stimulate the growth of roots and stems
- The tiny plantlets are transferred into potting trays where they develop into plants

PRODUCING CLONES

Nature has been cloning organisms for billions of years. For example, when a strawberry plant sends out a runner (a form of modified stem), a new plant grows where the runner takes root. That new plant is a clone. Similar cloning occurs in grass, potatoes and onions. People have been cloning plants in one way or another for thousands of years. For example, when you take a leaf cutting from a plant and grow it into a new plant (vegetative propagation), you are cloning the original plant because the new plant has the same genetic makeup as the donor plant. Vegetative propagation works because the end of the cutting forms a mass of non-specialised cells called a callus. With luck, the callus will grow, divide and form various specialised cells (roots, stems), eventually forming a new plant.

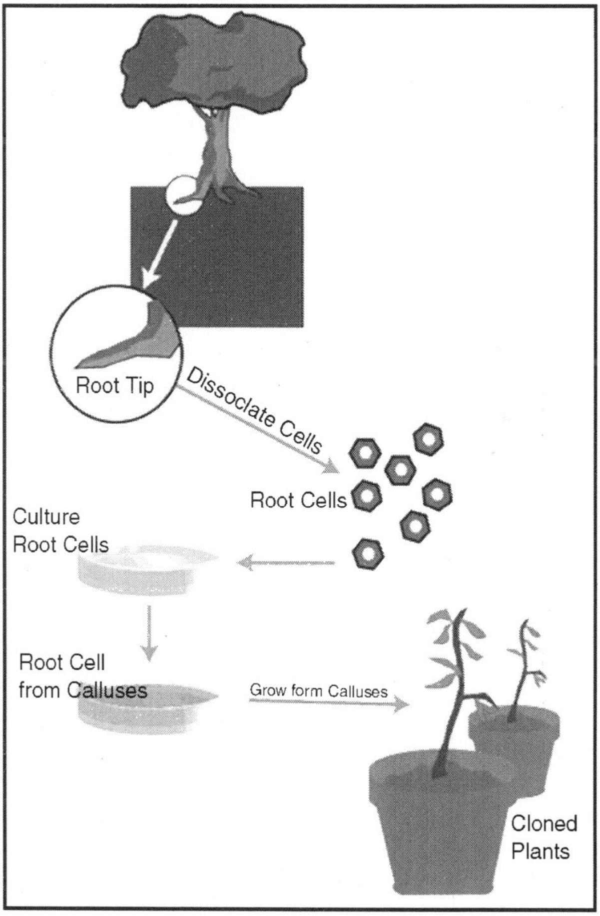

Fig. Diagram of Plant Cloning Through Tissue Culture Propagation.

More recently, scientists have been able to clone plants by taking pieces of specialised roots, breaking them up into root cells and growing the root cells in a nutrient-rich culture. In culture, the specialised cells become unspecialised (dedifferentiated) into calluses. The calluses can then be stimulated with the appropriate plant hormones to grow into new plants that are identical to the original plant from which the root pieces were taken. This procedure, called tissue culture propagation, has been widely used by horticulturists to grow prized orchids and other rare flowers. Plants are not the only organisms that can be cloned naturally. The unfertilized eggs of some animals (small invertebrates, worms, some species of fish, lizards and frogs) can develop into full-grown adults under certain environmental conditions — usually a chemical stimulus of some kind. This process is called parthenogenesis, and the offspring are clones of the females that laid the eggs.

Another example of natural cloning is identical twins. Although they are genetically different from their parents, identical twins are naturally occurring clones of each other. Scientists have experimented with animal cloning, but have never been able to stimulate a specialised (differentiated) cell to produce a new organism directly. Instead, they rely on transplanting the genetic information from a specialised cell into an unfertilized egg cell whose genetic information has been destroyed or physically removed.

In the 1970s, a scientist named John Gurdon successfully cloned tadpoles. He transplanted the nucleus from a specialised cell of one frog (B) into an unfertilized egg of another frog (A) in which the nucleus had been destroyed by ultraviolet light. The egg with the transplanted nucleus developed into a tadpole that was genetically identical to frog B.

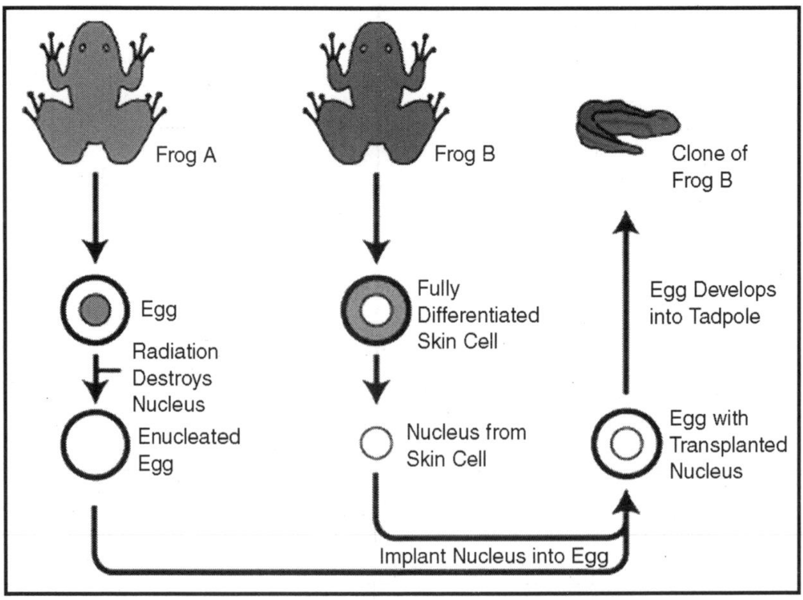

Fig. Gurdon's Experiment to Clone a Frog.

While Gurdon's tadpoles did not survive to grow into adult frogs, his experiment showed that the process of specialisation in animal cells was reversible, and his technique of nuclear transfer paved the way for later cloning successes.

Why Clone?

The main reason to clone plants or animals is to mass produce organisms with desired qualities, such as a prize-winning orchid or a genetically engineered animal — for instance, sheep have been engineered to produce human insulin. If you had to rely on sexual reproduction (breeding) alone to mass produce these animals, then you would run the risk of breeding out the desired traits because sexual reproduction reshuffles the genetic deck of cards.

Other reasons for cloning might include replacing lost or deceased family pets and repopulating endangered or even extinct species. Whatever the reasons, the new cloning technologies have sparked many ethical debates among scientists, politicians and the general public. Several governments have considered or enacted legislation to slow down, limit or ban cloning experiments outright. It is clear that cloning will be a part of our lives in the future, but the course of this technology has yet to be determined.

Who is it for ?

Therapeutic cloning can be used to fight diseases and create organs artificially.

Once specialised cells can be derived from cloned embryos the following diseases threatening millions of people on earth can be treated:

- Brain disorders like Parkinson and Alzheimer thanks to derived brain cells
- Diabetes thanks to derived insulin-producing pancreatic islet cells
- Spinal cord damage thanks to derived nerve cells
- Autoimmune disorders such as multiple sclerosis and rheumatoid arthritis thanks to derived cells of blood and bone marrow
- Cancer

In addition, it will help to save the lives of the thousands of people worldwide waiting for an organ transplantation by helping to create artificial organs. Not only is it difficult today to obtain an organ but even if organs are available many are being rejected by the body because of incompatability; let alone the brutal way some organs are being obtained, *i.e.*, by buying them from people in third world countries or even extracting them without the donor's consent !

Therapeutic cloning will eventually provide genetically 100 per cent compatible organs such as livers, kidneys, etc., and therefore save the lives of those who would perish without it.

Reproductive human cloning will help:

- *Infertile couples:* We have received many requests from infertile couples who cannot have children even after years of infertility treatments. For those people cloning is the only way to have a child of their own genetic offspring
- *Homosexuals:* They cannot have a child today that is 100 per cent related to them genetically but human cloning will provide this possibility for them.

TYPES OF CUTTINGS FOR CLONING PLANTS

Many resources on cloning plants sometimes refer to "plant cuttings," but don't explain the many ways that this can be done. There is real single way to cut plants. In fact, since virtually all of the cells in a plant can create every part that a plant needs to survive, a clone can be made from almost any of the plant's vegetation. Here are the four kinds of cuttings.

Stem Cutting

Just as its name implies, stem cutting involves taking a stem from the plant to make a clone. Stem cutting is perhaps the most popular cutting method for cloning plants in hydroponics. A stem cutting should be a couple inches in length and have a few healthy, large leaves to maximize photosynthesis. Make certain to keep the leaves of stem moist as it starts to take root.

Leaf Cutting

In this type of cutting for cloning plants, the leaf blade itself is used to create a new plant. Usually, the cut leaf doesn't become a part of the new plant. If the plant you are growing has particularly thick leaves, you can cut open the veins of the leaf and plant it flat into the growing medium, making sure to keep the cut leaf exposed to light and moisture. After a short period of time, new plants may begin to form where the leaf was cut open.

The cut leaf will eventually dry up and rot away, even as the young plants thrive. Another technique sometimes used is to take older, larger leaves and cut them into triangles, making sure each piece contains a large vein. The edges of the leaf, which usually contains no veins, should be thrown away. The triangles are then planted in the medium with the pointy side down. A new plant can then develop from the vein. Be aware, however, that leaf cutting is very delicate, and only works with plants that are easily cloned.

Leaf-Bud Cutting

This type of cutting is made up of a leaf blade, petiole, and a short piece of the stem with an attached bud. Leaf bud cuttings are best of species that are able to create roots but not shoots from cut leaves. This method works best when you have a healthy plant, but have very little cloning material to work with because it maximizes the propagating material.

Root Cutting

Propagation by cutting a parent's plant's roots is very simple, because there is less concern of the roots taking hold. However, it is not as popular because many gardeners don't want to harm the roots of their plants. Obviously, the thicker and healthier the root, the greater the chance of success. Roots should be two or three inches in length, and planted right side up in the growing medium.

Many growers choose to cut the side that is supposed to face upwards with a straight cut and the side that is supposed to face downwards with a diagonal cut so that they do not accidentally plant it wrong. No matter what kind of cutting you choose for cloning plants, make certain to use a quality rooting gel to stimulate the formation of new roots.

NATURAL CLONING

Clones occur naturally in plants. It can occur when plants such as the strawberry send out runners, which establish a whole new plant. If you plant an old potato it will grow into a clone of the original. Yet another example is plants such as daffodils, which produce bulbs. Quite often they split into two bulbs with each plant becoming a clone of the other. The cloning process occurs through cell division mechanism of mitosis. It therefore allows them to undergo this form of asexual reproduction. However, these plants can also reproduce using sexual reproduction (that is releasing gametes). This is important as it allows for genes to be shared between different individuals and then on to their offspring. This avoids the loss of genetic variation, which is the main problem of cloning.

ARTIFICIAL CLONING

For many years, gardeners have sat hunched in potting sheds around the world taking cuttings. A small piece of branch or stem is cut from a larger plant and is perhaps dipped into an auxin rooting powder. In a few weeks a new plant develops. In a few weeks a new plant develops. Little do these humble gardeners realise that they are carrying out a form of micropropagation. This is a high-tech version of the traditional cutting approach.

In micropropagation, cuttings are taken from a stem and cut into smaller sections. Each section is sterilised first before adding them to a growth medium containing rooting hormones. After each develops roots it grows into a plantlet. Finally, they are hardened up by being grown in a greenhouse. Tissue culture is another new technique that has been used for cloning plants. Here, only a few plant cells are needed. These are then added to the growth medium and hormones.

They will develop into a new plant:

- *The advantages:* The new systems are that they can produce new plants very quickly, in large numbers and in a small space. Also, they can grow inside all year round and within a controlled, disease-free environment.
- *The disadvantages:* Are that using cloned plants you face reducing the gene pool and therefore increasing the vulnerability to diseases.

ARTIFICIAL VEGETATIVE PROPAGATION OR CLONING

Vegetative propagation produces the next generation that is genetically identical to the parent. Such an organism that is genetically identical to the parent is called a clone. In case of plants with advantageous characteristics, the characteristics can be preserved by producing clones. This is particularly useful to agriculturists and horticulturists in order to get the best crop and uniform yield every time.

There are various ways of carrying out artificial propagation of plants. Cutting, layering, grafting and budding are some of the traditional methods whereas tissue culture is a recent technology.

Cutting

Cutting involves removing a piece of the parent plant - stem, root or leaf, and planting it in a suitable medium. At first roots are produced and then the shoot with the leaves. If a stem is taken, it must contain the nodal region. In some cases, rooting hormone may be required to initiate root formation.

For example:

- Stem cutting is commonly done for rose, sugarcane, banana, geranium, etc.,
- Root cutting is done for dahlia
- Leaf cuttings are used for African violets.

Layering

Layering is the method of inducing certain branches of the parent plant to produce roots by bending and pegging them to the ground around the parent plant leaving the tips exposed. Once the roots develop the branch is then cut off from the parent body. The branch that produces the roots is called the layer. It is a natural method in plants such as black raspberries. However, it is induced in plants like Jasminum, Rhododendron, strawberries, Magnolia, etc.

Grafting

It is the transfer of a part of one plant to the stump of another plant. The part taken from a plant is a portion of the stem with many buds. This portion is called scion and is selected for the quality of its fruit. The stump to which the scion is attached is called the stock. Stock is selected for qualities such as disease resistance and hardiness.

The cut ends of both the scion and stock are shaped such that they complement each other and their cambial tissues are close to each other. The two cut ends are brought together and covered with grafting wax. After some time, the tissues of the scion and the stock become continuous. The plant bears flowers and fruits characteristic of the scion. It is commonly practised on apple trees and on such plants which either do not produce viable seeds or the seeds which have a lot of variation.

Budding or Bud Grafting

It is a variation of the grafting method explained above. In this method, the scion is a bud along with some bark. A 'T'-shaped cut is made on the stock into which the scion is inserted and bound with a tape.

CLONING UNICELLULAR ORGANISMS

Cloning a cell means to derive a population of cells from a single cell. In the case of unicellular organisms such as bacteria and yeast, this process is remarkably simple and essentially only requires the inoculation of the appropriate medium. However, in the case of cell cultures from multi-cellular organisms, cell cloning is an arduous task as these cells will not readily grow in standard media.

A useful tissue culture technique used to clone distinct lineages of cell lines involves the use of cloning rings (cylinders). According to this technique, a single-cell suspension of cells that have been exposed to a mutagenic agent or drug used to driveselection is plated at high dilution to create isolated colonies; each arising from a single and potentially clonal distinct cell. At an early growth stage when colonies consist of only a few of cells, sterile polystyrene rings (cloning rings), which have been dipped in grease are placed over an individual colony and a small amount of trypsin is added. Cloned cells are collected from inside the ring and transferred to a new vessel for further growth.

Cloning Stem Cells

Somatic-cell nuclear transfer, known as SCNT, can also be used to create embryos for research or therapeutic purposes. The most likely purpose for this is to produce embryos for use in stem cell research. This process is also called "research cloning" or "therapeutic cloning." The goal is not to create cloned human beings (called "reproductive cloning"), but rather to harvest stem cells that can be used to study human development and to potentially treat disease. While a clonal human blastocyst has been created, stem cell lines are yet to be isolated from a clonal source.

Therapeutic cloning is achieved by creating embryonic stem cells in the hopes of treating diseases such as diabetes and Alzheimer's. The process begins by taking out the nucleus (containing the DNA) from an egg cell and putting in it a nucleus from the adult cell to be cloned. In the case of someone with Alzheimer's disease, the nucleus from a skin cell of that patient is placed into an empty egg. The reprogrammed cell begins to develop into an embryo because the egg reacts with the transferred nucleus. The embryo will become genetically identical to the patient. The embryo will then form a blastocyst which has the potential to form/become any cell in the body.

The reason why SCNT is used for cloning is because somatic cells can be easily acquired and cultured in the lab. This process can either add or delete specific genomes of farm animals. A key point to remember is that cloning is achieved when the oocyte maintains its normal functions and instead of using sperm and egg genomes to replicate, the oocyte is inserted into the donor's somatic cell nucleus. The oocyte will react on the somatic cell nucleus, the same way it would on sperm cells.

The process of cloning a particular farm animal using SCNT is relatively the same for all animals. The first step is to collect the somatic cells from the animal that will be cloned. The somatic cells could be used immediately or stored in the laboratory for later use. The hardest part of SCNT is removing maternal DNA from an oocyte at metaphase II. Once this has been done, the somatic nucleus can be inserted into an egg cytoplasm. This creates a one-cell embryo. The grouped somatic cell and egg cytoplasm are then introduced to an electrical current. This energy will hopefully allow the cloned embryo to begin development. The successfully developed embryos are then placed in surrogate recipients, such as a cow or sheep in the case of farm animals.

SCNT is seen as a good method for producing agriculture animals for food consumption. It successfully cloned sheep, cattle, goats, and pigs. Another benefit is SCNT is seen as a solution to clone endangered species that are on the verge of going extinct. However, stresses placed on both the egg cell and the introduced nucleus are enormous, leading to a high loss in resulting cells. For example, the cloned sheep Dolly was born after 277 eggs were used for SCNT, which created 29 viable embryos. Only three of these embryos survived until birth, and only one survived to adulthood. As the procedure currently cannot be automated, and has to be performed manually under amicroscope, SCNT is very resource intensive. The biochemistry involved in reprogramming the differentiated somatic cell nucleus and activating the recipient egg is also far from being well-understood.

In SCNT, not all of the donor cell's genetic information is transferred, as the donor cell's mitochondria that contain their own mitochondrial DNA are left behind. The resulting hybrid cells retain those mitochondrial structures which originally belonged to the egg. As a consequence, clones such as Dolly that are born from SCNT are not perfect copies of the donor of the nucleus.

CLONING BASED ON KNOWLEDGE OF PROTEIN STRUCTURE

Many of the first plant traits to be cloned were those in which some degree of biochemical understanding preceded any work with DNA. The understanding may have come from plant systems or from outside the plant kingdom.

Most of the cloned genes in these cases were enzymes which were part of a well-characterised biochemical pathway or had a protein structure which was otherwise well known, such as seed storage proteins. Cloning genes based on a knowledge of the protein sequence can be achieved by working backwards from the gene product (amino acid sequence) to the coding sequence or the DNA template. Such an approach has been used for cloning many seed storage proteins, such as the 2S sulphur- rich proteins from Brazil nut.

Genes of plant origin have been used in transgenic plants to improve resistance to insect pests. One of the first to be characterised was the cowpea trypsin inhibitor, a protease inhibitor which acts as an antifeedant to a range of insect pests. The protein is naturally produced in seeds of some genotypes of cowpea (*Vigna unguiculata*), which was shown to confer lepidopteran insect resistance when expressed constitutively in tobacco. Another such report from the same group at Durham

University described the use of a snowdrop lectin gene to confer aphid resistance in transgenic tobacco, the first report of genetically engineered resistance to a sucking insect in plants. The lectin from snowdrop (*Galanthus nivalis*), known as GNA, has been demonstrated to be effective against two major rice pests, the brown planthopper and the green leafhopper, as well as the peach potato aphid. When expressed at 0.1 per cent of total leaf protein in tobacco, GNA is antimetabolic to homopteran pests. These levels of protein were accumulated in leaf tissues when placed under the control of the CaMV 35S promoter. However, Hilder *et al.* (1995) postulate that better control would be achieved with the use of a phloem-specific promoter, such as the sucrose synthase 1 promoter from rice.

Another significant group of genes which have been cloned based on prior knowledge of the end product are the seed storage genes, including both seed storage proteins and lipids. In both cases there was considerable biochemical knowledge of the pathways involved in biosynthesis and storage. For seed storage proteins, the genes can be cloned based on the knowledge of the amino acid sequence of the mature protein and targeting signal or transit peptide. Molecular genetic manipulation of vegetable oils is a major area of research in the developed world. While approximately 90 per cent of vegetable oils are used for human consumption, there are emerging markets for industrial applications including lubricants, paints, cosmetics, plasticisers and soaps.

With increasing emphasis on quality and dietary considerations, there is a desire in developed countries to substantially alter the types and balances of fatty acids found in the major oil crops. Six crop species (soybean, oil palm, rapeseed/canola, sunflower, cottonseed and peanut) account for 84 per cent of the world's vegetable oil production, with the major fatty acids produced being palmitic, stearic, linoleic, linolenic and oleic acids.

However, almost 200 different fatty acids are produced by plants, mostly occurring in non-domesticated plant species. Hence, genetic manipulation of the major crop species, either via conventional breeding or genetic engineering, has the potential to modify the fatty acid types produced and stored by seeds such as soybean and canola. The biochemistry of fatty acid biosynthesis is well established for many of the major pathways, with enzymes identified which control the chain length of fatty acids, and the degree of saturation, or number of double bonds in the fatty acid chain. The oils stored in seeds have been altered by downregulation of particular enzymes using antisense RNA techniques.

The oleic acid content of rapeseed oil has been increased to 83 per cent (from 62 per cent in conventionally bred cultivars) by downregulating the enzyme D12-oleate desaturase using antisense RNA. A similar result has been achieved using co-suppression, which has produced rapeseed oil with 87 per cent oleic acid content. Genetic engineering approaches have also been used to introduce totally novel fatty acids to major oilseed crops. Petroselinic acid, an isomer of oleic acid, is found in abundance in the spice plant, coriander (*Coriandrum sativum*).

This oil has uses in cosmetics and pharmaceuticals, and its oxidation leads to the formation of lauric acid, used in detergents, and adipic acid, which is used in nylon production. Hence this is a potentially valuable oil which may enable soybean and canola crops – *i.e.*, renewable resources – to replace some of the non-renewable carbon fuels which are currently used for many industrial polymer and lubrication applications.

A cDNA clone from coriander encoding an acyl-ACP desaturase has been genetically engineered into tobacco, resulting in the accumulation of petroselinic acid. Oil crops such as rapeseed and soybean could be genetically modified to produce this fatty acid on a commercial scale. Hence, here is an example whereby a minor crop plant has become a donor of genetic diversity which will enable the commercial production of a very useful array of products for food and industrial purposes.

Cloning Genes Known Only by Phenotype

However, a more challenging task has been to move beyond proteins and pathways about which sufficient biochemistry is known into cloning genes known by phenotype only. By far the majority of genes fit into this category, which includes many of the most important plant traits such as disease resistance, adaptation to abiotic stresses, phenology and flowering behaviour and fertility. Only in the past few years have genes for traits known only by phenotype been cloned. One of the most interesting areas of plant biology is the understanding of plant–pathogen interactions at the molecular level.

The major methods used for cloning genes known only by phenotype are:

- Transposon tagging.
- Map-based or positional cloning.
- T-DNA tagging.

There have been a number of plant genes cloned using all these methods or modifications of them. There are also examples where a combination of methods has been used to expedite the rate of identification of the unknown sequence.

Transposon Tagging

The proposal that specific pieces of DNA are able to autonomously excise from a particular locus and insert, theoretically at random, into another linked or non-linked locus was first proposed by Barbara McClintock. Transposable elements are now accepted as ubiquitous, and are particularly well characterised in organisms such as *Escherichia coli, Drosophila* and maize. In two plant species, maize and the snapdragon (*Antirrhinum majus*), transposable elements are sufficiently well characterised to be used for mutagenesis and gene tagging.

The best understood system is the maize *Ac/Ds* (*Activator/Dissociation*) family of elements. The behaviour of the *Ac* element is such that it can excise autonomously from its locus, under the control of its own transposase, and insert into another chromosomal position. Where the *Ac* element is in or near a gene, that gene is inactivated (*i.e.*, not expressed), and will appear phenotypically as a recessive mutation. When *Ac* transposes (excises and reinserts at another locus), the gene is commonly returned to normal function, and hence is termed a revertant.

Should this *Ac* element then insert into another gene (transcribed or regulatory sequence), then a new recessive mutation will manifest. Transposition can occur either germinally, which results in an altered phenotype for the whole plant, or somatically, which results in a mosaic mutation, and can often be detected as a sectoring of plant or seed colour. The *Ds* element is actually a group of elements derived from defective *Ac* elements, with a common feature being their lack of effective *Ac* transposase. Hence these elements are non-autonomous, and will only transpose in the presence of an *Ac* element capable of producing transposase. Equipped with sequence and map information of an *Ac* element in maize, geneticists have been able to clone important maize genes using the transposon-tagging approach.

Table. Plant Transposable Elements which have been used for Transposon-Taggine

Element	Species	Type of Element	Size
Ac	Maize	Autonomous	4.6
Ds	Maize	Non-autonomous	0.5–30
Tam1	Snapdragon	Autonomous	15
Tam2	Snapdragon	Non-autonomous	5
En/Spm	Maize	Autonomous	8.3

When a gene is inactivated by an *Ac* element, proof of which can be demonstrated by a low frequency of trait reversion, the *Ac* element can be used as a probe for 'fishing-out' the tagged gene. A tagged gene can be cloned by performing Southern blot analysis with the *Ac* element used as probe, and any polymorphism represents an excision/insertion event. The technique is described more fully in the review by Walbot.

The *Ac* system has proven extremely useful for cloning maize genes for traits such as plant pigment genes (*bronze*; the anthocyanin pathway), starch biosynthesis (*opaque2*), and abscisic acid (AbA) insensitivity. Other maize transposable elements such as *Mutator* (*Mu*) and the *En/Spm* system have been used to clone maize genes, as has the *Tam* system in snapdragon. As a result, the search began for transposable element systems in other species so that a similar approach could be taken. However, it was soon demonstrated that the well-characterised *Ac/Ds* system could be genetically engineered into other plant species including tobacco, *Arabidopsis*, tomato and petunia. It was subsequently shown that the *Ac* element acted in a similar manner in these heterologous species, undergoing germinal and somatic excision and insertion, and maintaining the propensity to reinsert at a linked site (usually within 4 cM). As a result, the frequency of mutation in the surrounding area approaches 1024. It was not until 1993 that a plant gene was cloned using the heterologous system, that of the *Ph6* gene in petunia, which is involved in floral pigmentation. Using the transposon-tagging approach, a number of plant disease resistance genes (R genes) have now been cloned. The first of these was the tomato *Cf-9* gene, which confers resistance to race 9 of the fungal pathogen *Cladosporium fulvum*, causal agent of leaf mould. The gene was tagged using a modified *Ac/Ds* system.

The system relied on the use of a stabilised *Ac* element (s*Ac*), which was effectively a disabled *Ac* element that nevertheless produced *Ac* transposase and hence enables *Ds* transposition. The two elements are maintained in separate lines, and hence are both stable until controlled crossing is performed. This gives a significant advantage in that *Ds* insertion mutants can be maintained indefinitely in the homozygous or hemizygous form, and revertants can be isolated by crossing lines with s*Ac* tomatoes. The transposon-tagging approach used to isolate the *Cf-9* gene, which was known by phenotype only, did have considerable advantages in that much was known of the pathogen. The hypersensitive response elicited by the virulent strain fitted into the gene-for-gene hypersensitive response proposed by Flor, with a dominant avirulence gene (*Avr9*) interacting with the *Cf-9* gene to form an incompatible or hypersensitive response. *Avr*9 had previously been cloned and fully characterised, and was known to specify a 28 amino acid peptide which alone could elicit a necrotic response in resistant tomatoes. Crosses were made to develop a transgenic tomato line homozygous for *Cf-9*, but heterozygous for *Ds*. Crossing was performed as outlined in Figure.

As can be seen, most seedlings died from systemic hypersensitive response, with *Avr9* and *Cf-9* present in all cells. The only survivors would be individuals where *Ds* had inserted into the *Cf-9* gene. Where this happened in plants carrying s*Ac*, the seedlings were variegated for the hypersensitive response (somatic excision), and in the absence of s*Ac*, the survivors were stable (germinal excision).

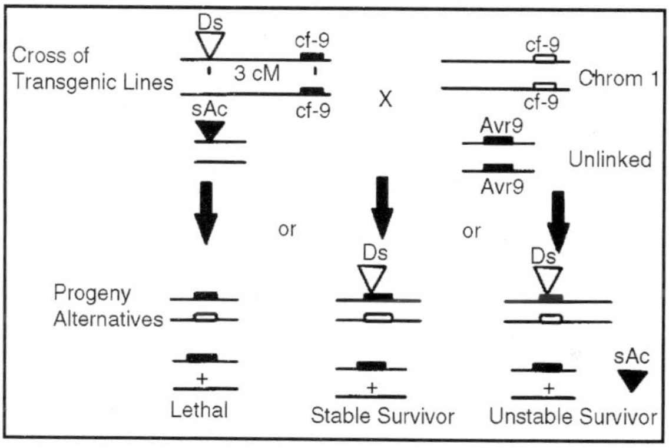

Fig. Graphical Representation of the Crosing Strategy of Transgenic Ac/Ds Tomato Lines.

This work was a major undertaking, involving the germination of approximately 160 000 progeny, yielding 118 survivors, representing 63 independent mutations. After considerable effort to characterise these, the *Cf-9* gene was cloned and characterised, with an open reading frame encoding an 863 amino acid protein.

Map-based Cloning

The previously described transposon-tagging scheme utilised to clone *Cf-9* had an element of the map-based approach by utilising the phenomenon of *Ac*/*Ds* transposition to linked sites, with a *Ds* element within 3 cM of the locus of interest, which greatly enhanced the likelihood of insertional mutagenesis. However, map-based cloning requires a much tighter genetic linkage (in centimorgans or recombination fraction), which in turn needs to be converted into a physical distance (kbp of DNA sequence). The map-based, or positional cloning concept relies on building a well-saturated genetic map of DNA sequences of the organism in question. The genetic map is built up in a polymorphic mapping population, commonly an F2, backcross (BC) or recombinant inbred, with markers such as RFLPs (restriction fragment length polymorphisms). The first RFLP map to be made was of human. Based on such maps, important human genes responsible for disorders such as cystic fibrosis and muscular dystrophy have been cloned in this manner. Genetic maps of plant species have been made with DNA markers in virtually all important plant species, particularly the major cereals, grain legumes, oilseeds and important vegetable crops including tomato and potato. The RFLP marker system is based on Southern blot analysis, while many of the newer technologies use the polymerase chain reaction (PCR).

These include random amplified polymorphic DNA (RAPDs), amplified fragment length polymorphisms (AFLPs), and a number of different marker systems based on simple sequence repeats (SSRs), which may use PCR, such as inter simple sequence repeat (ISSR)-PCR, or Southern analysis. The applications of DNA markers to plant genetic analysis and breeding are many, and the subject has been reviewed by various authors. The population for saturation mapping should be segregating in a Mendelian manner for the trait of interest. This requires a lot of tedious but demanding work, and it may be that particular populations such as near isogenic lines, or DNA pooling strategies, such as bulked segregant analysis, will allow the mapping effort to be more targeted to the region of interest, hence improving the rate of map saturation. It is desirable to obtain extremely close genetic linkage to the locus of interest (<1 cM), with flanking markers if possible. It can be of considerable advantage to identify a marker which co-segregates (say within 0.1 cM) with the trait. The next step is to convert the genetic map into a physical map of the chromosome region, translating distances from centimorgans to nucleotides. The Gene of Interest (in this illustration,R) is Flanked by Markers such as RFLPs or SSRs. A Series of Overlapping YAC Contigs is used to cover th Chromosomal Region Containing the Gene.The YAC Which Contains the Complete Gene is Marked in Bold.This is then Subcloned or Matched to a cDNA Library to Idenfify the Gene(s) Contained the Region, and Ultimately Identify the R Gene.

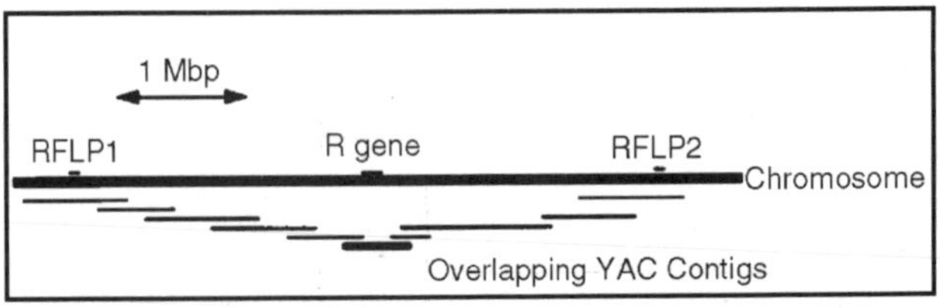

Fig. Generalised Strategy of Chromosome Walking Used in Map-based gene Cloning.

Physical mapping then involves the use of YAC or BAC clones which span the region of interest. These YAC/BAC contigs will span the region between two flanking DNA marker sequences as shown in Figure, a process known as chromosome walking. In this example, two YAC/BAC clones contain the gene of interest. Methodology is being developed to enable BAC clones to be directly transformed into plants, to find which one contains the gene of interest (in the example, the *R* gene). Alternatively, the YAC/BAC must be subcloned (for example as a cDNA library), and these subclones can be put into the susceptible plant and tested for the R phenotype. The first R gene to be cloned in plants via map-based cloning was the *Pto* gene, which confers resistance to *Pseudomonas syringae* pv. *lycopersici*, the causal agent of bacterial speck in tomato.

The starting points for the work were:

- A high-density RFLP map of tomato.
- A tomato YAC library.

The *Pto* locus had been previously mapped to tomato chromosome 5 and one RFLP marker co-segregated with *Pto*. Hence this was a useful starting point to screen the YAC library. A 400 kb YAC was identified, and via inverse PCR, some end-specific probes were generated. The left arm of the YAC was 1.8 cM from *Pto*, with the right arm perfectly co-segregating. This YAC was then used to screen a 920 000 plaque leaf cDNA library, from which 30 were selected. One of these was shown to co-segregate with *Pto*. This cDNA was then transformed into a susceptible tomato cultivar under the control of the CaMV 35S promoter, where it conferred resistance. Hence the gene was identified, and could then be sequenced and characterised.

The success of this approach was partly due to the small genome size of tomato (950 Mbp), and a relatively low level of repeat sequence. Such an approach is much more difficult in species with larger genomes such as maize (2300–2700 Mbp) and wheat (16 000 Mbp). One such example of this problem is the 79 000 clone maize YAC library produced by Zeneca Seeds. As reported by K. Edwards, only 15 per cent of the YACs have unique ends, with the remainder terminating in repeat sequences, which makes chromosome walking very difficult. The synteny of genomes is a major advantage in this instance. It is now well established that there is a high degree of synteny of genome organisation, as seen in the co-linearity of maps of the most important grass genomes. This has resulted in the great interest in using this co-linearity to facilitate cloning of genes in large genomes such as maize and wheat, by starting with the smaller genomes, particularly rice, and to a lesser extent, sorghum.

One particular study of maize YACs demonstrated that a single YAC with the *Adh1* sequence contained 36 different repetitive sequences, whereas these repetitive sequences were largely absent in corresponding areas of the rice and sorghum genomes, a pattern seen also for the *a1–sh2* region in all three genomes. In a similar manner, many genes of interest for *Brassica* improvement are being identified in the smaller *Arabidopsis thaliana* genome. Paterson *et al.* have recently demonstrated the potential of this approach, by identifying quantitative trait loci (QTLs) for seed mass and phenology which were co-linear on molecular marker maps of maize, sorghum and rice.

They proposed that, as the maize genome is four and six times larger than the sorghum and rice genomes, respectively, the genes of interest in maize may be cloned in sorghum/rice more expediently. It must be remembered that QTLs are subject to genotype 3 environment interaction, which will be a major complicating factor in attempts to clone and characterise a QTL.

RECOMBINANT DNA AND GENE CLONING

Recombinant DNA is DNA that has been created artificially. DNA from two or more sources is incorporated into a single recombinant molecule.

Making Recombinant DNA (rDNA)

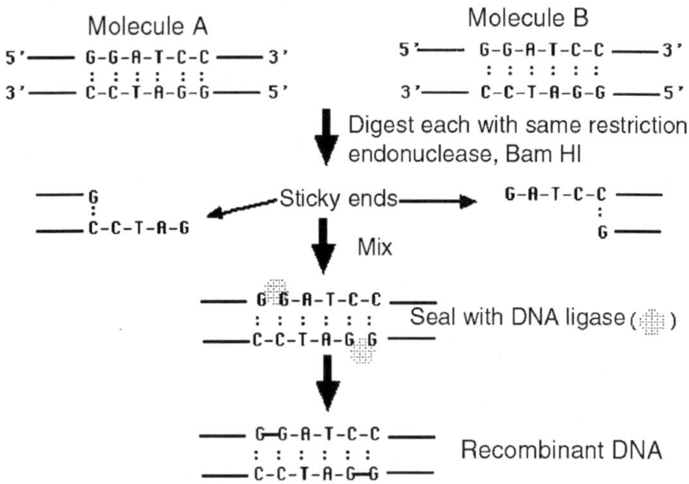

- Treat DNA from both sources with the same restriction endonuclease (BamHI in this case).
- BamHI cuts the same site on both molecules
  5' GGATCC 3'
  3' CCTAGG 5'
- The ends of the cut have an overhanging piece of single-stranded DNA.
- These are called "sticky ends" because they are able to base pair with any DNA molecule containing the complementary sticky end.
- In this case, both DNA preparations have complementary sticky ends and thus can pair with each other when mixed.

- DNA ligase covalently links the two into a molecule of recombinant DNA.

To be useful, the recombinant molecule must be replicated many times to provide material for analysis, sequencing, etc. Producing many identical copies of the same recombinant molecule is called cloning. Cloning can be done in vitro, by a process called the polymerase chain reaction (PCR). Here, however, we shall examine how cloning is done in vivo.

Cloning in vivo can be done in

- Unicellular microbes like E. coli
- Unicellular eukaryotes like yeast and
- in mammalian cells grown in tissue culture.

In every case, the recombinant DNA must be taken up by the cell in a form in which it can be replicated and expressed. This is achieved by incorporating the DNA in a vector. A number of viruses (both bacterial and of mammalian cells) can serve as vectors. But here let us examine an example of cloning using E. coli as the host and a plasmid as the vector.

PLASMID

A plasmid is an extra-chromosomal DNA molecule separate from the chromosomal DNA which is capable of replicating independently of the chromosomal DNA. In many cases, it is circular and double-stranded. Plasmids usually occur naturally in bacteria, but are sometimes found in eukaryotic organisms (*e.g.*, the 2-micrometre-ring in Saccharomyces cerevisiae).

Plasmid size varies from 1 to over 200 kilobase pairs (kbp). The number of identical plasmids within a single cell can be zero, one, or even thousands under some circumstances. Plasmids can be considered to be part of the mobilome, since they are often associated with conjugation, a mechanism of horizontal gene transfer.The term plasmid was first introduced by the American molecular biologist Joshua Lederberg in 1952.

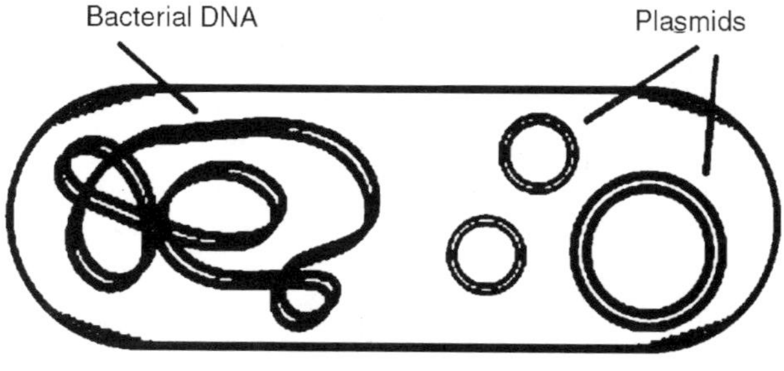

Fig: Plasmids Enclosed Showing Chromosomal DNA and Plasmids.

Plasmids can be considered to be independent life-forms similar to viruses, since both are capable of autonomous replication in suitable (host) environments. However the plasmid-host relationship tends to be more symbiotic than parasitic (although this can also occur for viruses, for example with Endoviruses) since plasmids can endow their hosts with useful packages of DNA to assist mutual survival in times of severe stress. For example, plasmids can convey antibiotic resistance to host bacteria, who may then survive along with their life-saving guests who are carried along into future host generations.

Vectors

There are two types of plasmid integration into a host bacteria: Non-integrating plasmids replicate as in the top instance; whereas episomes, the lower example, integrate into the host chromosome. Plasmids used in genetic engineering are called vectors. Plasmids serve as important tools in genetics and biotechnology labs, where they are commonly used to multiply (make many copies of) or express particular genes. Many plasmids are commercially available for such uses. The gene to be replicated is inserted into copies of a plasmid containing genes that make cells resistant to particular antibiotics and a multiple cloning site (MCS, or polylinker), which is a short region containing several commonly used restriction sites allowing the easy insertion of DNA fragments at this location.

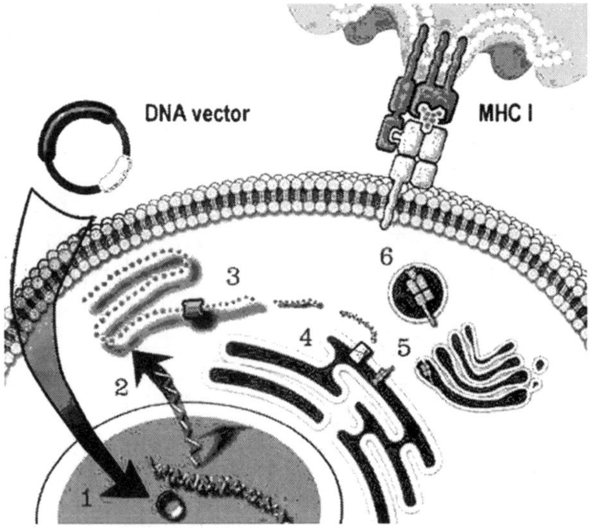

Fig: Plasmid Vector

Next, the plasmids are inserted into bacteria by a process called transformation. Then, the bacteria are exposed to the particular antibiotics. Only bacteria which take up copies of the plasmid survive the antibiotic, since the plasmid makes them resistant. In particular, the protecting genes are expressed (used to make a protein) and the expressed protein breaks down the antibiotics. In this way the antibiotics act as a filter to select only the modified bacteria. Now these bacteria can be grown in large amounts, harvested and lysed (often using the alkaline lysis method) to isolate the plasmid of interest.

Another major use of plasmids is to make large amounts of proteins. In this case, researchers grow bacteria containing a plasmid harboring the gene of interest. Just as the bacteria produces proteins to confer its antibiotic resistance, it can also be induced to produce large amounts of proteins from the inserted gene. This is a cheap and easy way of mass-producing a gene or the protein it then codes for, for example, insulin or even antibiotics. However, a plasmid can only contain inserts of about 1-10 kbp. To clone longer lengths of DNA, lambda phage with lysogeny genes deleted, cosmids, bacterial artificial chromosomes or yeast artificial chromosomes could be used.

TYPES

One way of grouping plasmids is by their ability to transfer to other bacteria. Conjugative plasmids contain so-called tra-genes, which perform the complex process of conjugation, the transfer of plasmids to another bacterium. Non-conjugative plasmids are incapable of initiating conjugation, hence they can only be transferred with the assistance of conjugative plasmids, by 'accident'. An intermediate class of plasmids are mobilizable, and carry only a subset of the genes required for transfer. They can 'parasitize' a conjugative plasmid, transferring at high frequency only in its presence. Plasmids are now being used to manipulate DNA and may possibly be a tool for curing many diseases.

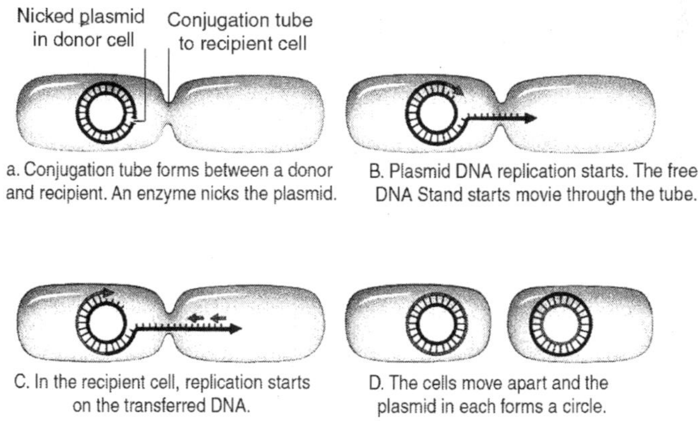

Fig: Bacterial conjugation

It is possible for plasmids of different types to coexist in a single cell. Seven different plasmids have been found in E. coli. But related plasmids are often incompatible, in the sense that only one of them survives in the cell line, due to the regulation of vital plasmid functions. Therefore, plasmids can be assigned into compatibility groups. Another way to classify plasmids is by function. There are five main classes:

- Fertility-F-plasmids, which contain tra-genes. They are capable of conjugation.
- Resistance-(R)plasmids, which contain genes that can build a resistance against antibiotics or poisons. Historically known as R-factors, before the nature of plasmids was understood.
- Col-plasmids, which contain genes that code for (determine the production of) bacteriocins, proteins that can kill other bacteria.
- Degradative plasmids, which enable the digestion of unusual substances, *e.g.*, toluene or salicylic acid.
- Virulence plasmids, which turn the bacterium into a pathogen.

Plasmids can belong to more than one of these functional groups. Plasmids that exist only as one or a few copies in each bacterium are, upon cell division, in danger of being lost in one of the segregating bacteria. Such single-copy plasmids have systems which attempt to actively distribute a copy to both daughter cells.

Some plasmids include an addiction system or "postsegregational killing system (PSK)", such as the hok/sok (host killing/suppressor of killing) system of plasmid R1 in Escherichia coli. They produce both a long-lived poison and a short-lived antidote. Daughter cells that retain a copy of the plasmid survive, while a daughter cell that fails to inherit the plasmid dies or suffers a reduced growth-rate because of the lingering poison from the parent cell.

Plasmid DNA Extraction

As alluded to above, plasmids are often used to purify a specific sequence, since they can easily be purified away from the rest of the genome. For their use as vectors, and for molecular cloning, plasmids often need to be isolated.

There are several methods to isolate plasmid DNA from bacteria, the archetypes of which are the miniprep and the maxiprep/bulkprep. The former can be used to quickly find out whether the plasmid is correct in any of several bacterial clones. The yield is a small amount of impure plasmid DNA, which is sufficient for analysis by restriction digest and for some cloning techniques.

In the latter, much larger volumes of bacterial suspension are grown from which a maxi-prep can be performed. Essentially this is a scaled-up miniprep followed by additional purification. This results in relatively large amounts (several micrograms) of very pure plasmid DNA.

9

Gene Expression in Plants

NEW CHALLENGE OF GENE TECHNOLOGY

With few exceptions classical bioprospecting for drugs has not proved economically beneficial for developing nations, but nor has it directly damaged these economies. In contrast, bioprospecting for genes may soon pose a real threat to the economic survival of these biodiversity-rich countries. The farmers of the North are currently suffering under low prices and overproduction of certain traditional crops and livestock. As a result they are looking for new markets and products.

Modern biotechnology promises to aid Northern farmers in this endeavour, eliminating or displacing traditional export commodities from developing countries and transferring production or substitutes from the farm fields of the South to those of the North. Quite possibly the transfer may even skip the Northern farms and jump straight into bioreactors. In Africa alone, US$ 10 billion in exports are vulnerable to industry-induced changes in raw material prices and requirements.

Most developments in plant biotechnology have been achieved with crops cultivated mainly in industrialized countries. This denies the farmers of developing countries the chance to benefit from the new agricultural opportunities of gene technology. As mentioned above, Calgene's high-laurate canola oil may displace coconut and palm kernel oil, posing a threat to the economic survival of millions of farm families in the South. In the Golfito region of Costa Rica the government has started a programme to grow oil palms on banana fields deserted by the US fruit multinationals. All these efforts, which are partially financed with Northern developmental aid, may vanish into thin air as a result of Calgene's new rape seed.

Thaumatin, a sweet-tasting basic protein from the tropical plant *Thaumatococcus*, has been traditionally used in West Africa as a sweetener. With the collection of *Thaumatococcus* fruits for the British food industry, the population in this region earned a large part of its income. However, the thaumatin gene has now been cloned and the sweetening protein can be produced by large-scale fermentation of brewer's yeast at low cost. The same applies to natural compounds like indigo, which can now be produced by the fermentation of *Escherichia coli* engineered with genes from a toluol-degrading subspecies of the soil bacterium *Pseudomonas putida*. Products like vanilla, pyrethrum and rubber may follow this same path.

Along the same lines, the extension of patent laws to the developing nations through the GATT could mean that the biotech industry obtains a monopoly on genetically engineered livestock and crops, which farmers in developing countries must cultivate under constraint in order to remain competitive. For example, US-Patent No. 5 159 135 of Agracetus covers all genetically engineered cotton. This patent is a warning of potential future problems, and has already caused an outcry in developing countries like India. If the biotech industry continues developing without adequate controls, the consumers and farmers of developing countries may even end up paying royalties on biotech products that were originally developed from their very own resources and knowledge.

As a step in the right direction, the US Patent and Trademark Office reversed its decision to grant the Agracetus patent at the end of 1994, but primarily as a result of pressure from the US biotech industry which argued that patents like this will inhibit research and development. Such issues bring important questions to light: How can developing nations be motivated to conserve their biodiversity under these threatening circumstances?

Are the regulations and tools of modern bioprospecting, as described above, sufficient to face this challenge? The Biodiversity Convention attempts to address this new threat by requiring that access to biodiversity's genetic potential be combined with the biotechnology transfer to the South in order for those countries to develop their own methods of sustainable biodiversity utilization. Nevertheless, the Convention suffers from three major drawbacks.

- The treaty was not ratified by the USA, the leading country in biotechnological research, development and application.
- It specifically excludes (under US pressure) *ex situ* genebank material collected before the enactment of the treaty. As a result, huge stocks of germplasm collected by the North, mostly in tropical and subtropical countries, are not restricted by the Convention. The recent transfer by the Consultative Group on International Agricultural Research of its 12 genebanks to the auspices of the UN must be the first step in keeping the access of developing countries open to their own resources.
- It still remains nearly impossible to control the illegal transfer of genetic material into the North. Genes can be cloned from minute amounts of DNA or RNA and isolated from biological material that easily fits into an airmail envelope. Genes do not have tags designating their country of origin, and once they are cloned, they

are no longer controlled by their source country. This is quite different from the isolation of natural compounds from plants, where larger amounts of plant material must be collected and, for the process of isolation and structure elucidation, must be re-collected. In this last case, controlling the flow of biological material is possible simply because industry will eventually require sample resupply at a given point, and for this the industry needs reliable partners in countries of origin.

These issues must be approached in an active and more aggressive manner than traditionally used. During collaborations with traditional pharmaceutical and biotech companies like Merck & Co., Bristol-Myers Squibb and the British Technology Group, INBio used such an approach, proving to industry that fair partnerships are mandatory and conducive to success. More importantly, INBio demonstrated that reliable applied research is possible in a developing country, and that technology transfer to acquire necessary know-how and equipment works to the advantage of the industrial partner as well. The same applies to the biotech industry.

In this case, the transfer of gene technology is not critical, in contrast to natural compound chemistry, because it does not require million-dollar investments in laboratory infrastructure. Rather, the industry relies on the know-how already existing in many source countries. Gene technology also represents a promising development opportunity for countries that do not have large research budgets at their disposal. Moreover, gene technology is a very straightforward approach, relying on natural history observations (*e.g.* whether certain plants show natural resistance to pathogens), and not involving the automated random screening of thousands of samples. Biodiversity inventories, which are already in place in countries like Costa Rica, are a reasonable and advantageous prerequisite for successful 'gene prospecting'.

Costa Rica's aggressive strategy to foster collaborations with the international industry and academic institutions in drug research, gene technology and agriculture actively seeks to develop and patent natural compounds, proteins and genes in Costa Rica based on a foundation of national research. Collaborations of this nature will help launch Costa Rica onto a scientific and technological plane that offers services and goods that are both competitive and compatible with those of industrial nations. Simultaneously, INBio works within this strategy to increase knowledge about Costa Rican biodiversity in general, access revenues for further conservation efforts, and to assign biodiversity a higher value than it has had in the past. There is little doubt that such activities will encourage society's willingness to preserve biodiversity for future generations, by making it worthwhile for the Costa Rican population to maintain tropical forests and other ecosystems on their own.

IMPORTANT OF GENE MAPPING

An important, though by no means an essential, step in genetic analysis is to produce genetic maps of the marker loci. Such maps are often referred to as 'framework maps' because they provide a framework within which important genes can be located, as well as providing a means of comparing chromosome organization in other closely or distantly related species. There are two stages to mapping. Firstly, to arrange the markers in a linear sequence separated by an appropriate map distance, *i.e.* to construct a linkage map. The second is to relate the linkage maps to particular recognizable chromosomes. The latter is often the most difficult and generally not essential for breeding or conservation work. We will, therefore, concentrate on the former.

Chromosomes contain a single linear molecule of DNA and hence the markers on that chromosome occur at particular positions along that molecule. Typically, each chromosome contains 107 to 108 base pairs (bp), *i.e.* 104 to 105 kilo-base pairs (kbp) of DNA, while a typical structural gene, coding for a polypeptide chain, would be between 1 and 2 kbp long. Assuming only 10% of the genome is coding DNA, then a chromosome probably contains something of the order of 1000 to 10 000 genes. Fortunately, the fact that the chromosome is a linear molecule means that the genes need to be mapped in only one dimension. Currently, the only useful method of gene mapping, at least as far as breeders and conservationists are concerned, relies on recombination between homologous chromosomes resulting from genetic exchange, which can be seen as chiasmata at diplotene through first metaphase of meiosis. A single chiasma on a particular chromosome results in half the gametes from such a meiosis being recombinant for that chromosome: half the gametes will contain a copy of that chromosome that is part maternal and part paternal in origin, while the remaining gametes will be entirely parental. Very few plant species have a sufficiently low number of clearly recognizable chromosomes that the number of chiasmata on a particular chromosome can be compared in different nuclei.

However, in many species the total number of chiasmata in each diplotene nucleus can be counted, at least in pollen meiosis, and the average number of chiasmata per nucleus calculated. The number and position of chiasmata on a particular pair of homologous chromosomes will vary from nucleus to nucleus; it is as though there are a large number of potential sites of exchange, but that in any given meiosis only a few sites are actually involved.

The longer the chromosome the more potential sites there might be. There will invariably be one chiasma and there may be two, three, four or more; a typical set of results is shown in Table 8.1a. A chromosome that has one chiasma on average is said to be 50 centimorgans (cM) long – a map length unit named after the American geneticist Thomas Hunt Morgan. This relates to the fact mentioned above that one chiasma results in 50% recombinant chromosomes. By extension, a chromosome with an average of 2.5 chiasmata is said to be 125 cM (*i.e.* 2.5 350 cM) long. It appears that, as a rough rule of thumb, each chromosome has on average two chiasmata and so is 100 cM long. It follows, therefore, that providing the haploid chromosome number is known (n), the total map length will be approximately 2 3 n 3 50 cM. The average number of chiasmata per nucleus has been calculated in many species and this rule generally holds.

Table. Chiasma Fraquency and Mapping.Chiasma Frequency in Chromosomes of Secale Cereale.

Number of chiasmata per chromosome	Frequency	Percentage unrecombined Chromosomes
1	0.267	50.0
2	0.716	25.0
≥3	0.017	12.5
Mean	–	22.0

This rough rule is important because it gives the geneticist an idea of the total genetic map length that should be expected as more genes or markers are mapped. It provides a guide to the extent to which the currently mapped markers cover the full genetic map. Because the map is based in chiasma units, it is not the same as a physical map. The distribution of chiasmata is not uniform over a given chromosome in a species and will vary with the chromosome and the species. Although a knowledge of the number of chromosomes and, ideally, the mean chiasma frequency per nucleus for the species, indicates the total map length, the actual map has to be constructed from examining the frequencies of progeny in crosses; *i.e.* maps are constructed from information based on the consequences of chiasmata, namely recombination of genetic markers. Recombination of genetic markers is identified by the presence of gametes that contain recombined markers, and such gametes can be recognized from the phenotypes of progeny of a cross.

The frequency of such gametes is an estimate of the recombination frequency between the two markers concerned. The closer two markers are to each other on the chromosome, the less likely a chiasma will occur between them and the less likely they are to recombine. Providing the two markers are sufficiently close that either none or one chiasma occurs between them but never more, then the recombination frequency as a percentage is equal to the map distance in centimorgans as described above. The problem arises when the markers are sufficiently far apart for two or more chiasmata to occur between them in some meioses. It can be shown that not only does a single chiasma between a pair of markers result in 50% recombination between them, but so also do two, three or more chiasmata, on average.

It is necessary to say 'on average', because there are a variety of possible consequences with two, three or more chiasmata in any given meiosis but, in practice, when one is looking at the progeny of a cross or self, each progeny will be the result of gametes from different meioses. This fact implies that although map distance increases linearly with the number of chiasmata, the same is not true with the frequency of recombination, which can never be more than 50% for a pair of markers even though they might be 100 cM apart at opposite ends of a chromosome.

These relationships are illustrated in Figure. In order to overcome this problem, mapping functions have been devised to correct for multiple chiasmata. The most common of these are the Haldane (1919) and Kosambi (1943) mapping functions. Haldane's assumes that the probability of none, one, two or more chiasmata in a given interval follows a Poisson distribution, *i.e.* that the chiasmata are independent, random events. Kosambi's method allows for a certain degree of nonindependence in chiasma occurrences. It has long been known by cytogeneticists that chiasma interference occurs over quite large regions of a chromosome.

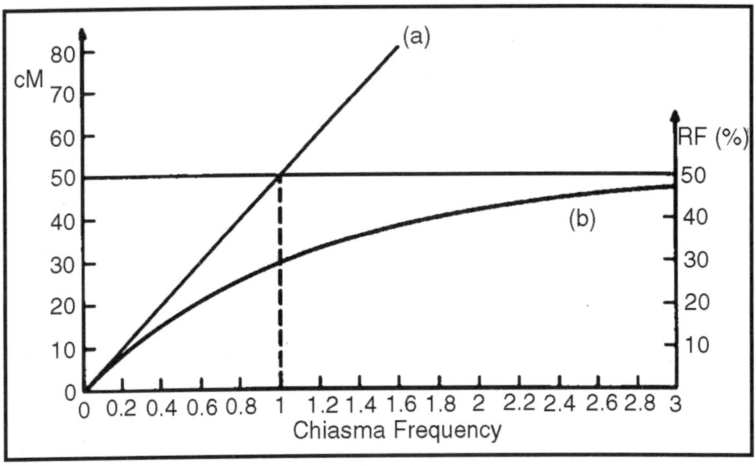

Fig. The Relationship of Chiasma Frequency With (a) Genetic Map Length and(b)Recombination Frequency.

Interference means that if a chiasma occurs at a particular point on the chromosome, the next one will never occur closer than some fixed distance, the interference distance, from it. Beyond this distance, there is a short distance in which the interference disappears and the next chiasma can then occur randomly outside this range. Observational data in various species of plants are confirming this and suggest that the interference distance is generally between 15 and 20 cM, *i.e.* 15 to 20% of a typical chromosome on either side of the chiasma. This has important and useful consequences for genetic and breeding work as will be shown later.

Recombination frequencies between pairs of markers can be scored in a wide variety of different crosses. The simplest to use are generations derived from an F1 because only two alleles are segregating, and their distribution in the chromosomes of the parents of the F1 can be determined. The generations that can be used are F2, backcrosses (Bc), recombinant inbred lines (RILs) or doubled haploid (DH) lines, with F2s being the most informative. General formulae for calculating recombination frequencies are given by Allard (1956). In outbreeding species where F1s are not available, a given individual can be considered as an F1 and its selfed progeny as being an F2, even though the inbred parents do not exist. If it cannot be selfed but controlled crossing to another individual is possible, then again those genes segregating in the cross can be mapped although the situation is more complex.

There could be from two to four alleles segregating at each locus and, with respect to any one locus, the cross could represent an F2, Bc1 or Bc2. Moreover, the distribution of alleles in the two parents cannot be known and has to be inferred from the progeny. This complexity is compounded by the large number of marker loci that may be segregating in any given cross. Fortunately, a range of software packages are available to estimate these recombination frequencies, identify linkage groups, assign the markers to the most likely order and space them in map units (cM) on these linkage groups; examples are MAPMAKER and JOINMAP. Ideally the number of linkage groups should be equal to the chromosome number in the gametes but, unless the markers available provide a good coverage of the genome, it is likely that those markers on a given chromosome may appear as two or more separate linkage groups simply because the subsets are not sufficiently close for them to be recognized as being together on one chromosome.

A typical marker framework map is shown in Fig. for the chromosomes of *Brassica oleracea*. Clearly these maps provide a very detailed coverage of the chromosome. Many of the markers are very close, *i.e.* less than 10 cM, and over this distance it matters little which of the two mapping functions is used. With the more widely spaced markers, *i.e.* recombination frequency >15%, Haldane's function will exaggerate their distance apart because it will allow for more double crossovers than actually occur. As more markers are mapped, the total length should converge on the value predicted by the chiasma frequency, *i.e.* approximately 2 3 *n* 3 50 cM. It is important to remember that genetic map distances have standard errors which are dependent on the size and type of population used to construct the map, the mapping population. The map obtained is that which best fits, given the data and the assumptions underlying it, but there may well be other maps which also fit the data well, though are slightly less likely.

Any particular map always develops a greater aura of respectability once it is published, frequently without stating its reliability, and subsequent yet different maps are treated with suspicion. A recombination frequency of true value p has a standard error of ?[p(1 2 p)/N], where N is the size of the gamete population. For example, if p is 0.1 (*i.e.* RF = 10%) and$N$ = 100, the standard error is 0.03 or 3%. In other words the 95% confidence interval of the estimate lies approximately between 4% and 16%. There are other reasons, apart from statistical sampling, why maps could be wrong. The most obvious is the accuracy with which the data are scored and recorded. Autoradiographs and banding patterns on gels can easily be misread and research workers are loath to discard data even though its interpretation is ambiguous. All data should be checked independently by two or more people and any ambiguities removed. Wrong scores will bias the data and often exaggerate map lengths as they suggest double recombination; wherever double recombination appears to have occurred in two short, adjacent intervals, the data should be checked. Changes in methylation rather than scoring errors could cause a restriction site to appear or disappear so giving the false impression of double crossing- over, or it could simply be an error.

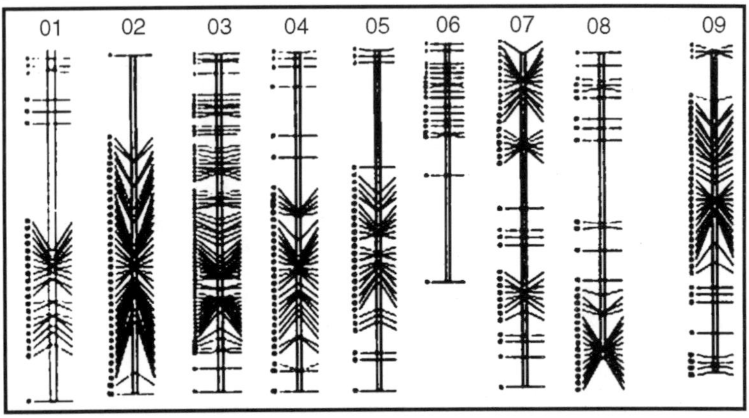

Fig. Typical Molecular Marker Framework Map.The Nine Groups of Brassica Oleracea Based on RFLP, Isozyme and Morphological Markers

Where maps appear to have major inconsistencies in different crosses, different chromosomal structural arrangements may be responsible, such as translocations or inversions. There is considerable evidence that chiasma frequencies, and hence recombination frequencies, may be quite different in male and female meioses even on the same plant although in other studies this may not be so. Thus although the order of markers on a linkage group may remain the same, the relative distances between them may be quite different if the map derives from male versus female meioses. For example, in *Brassica* it appears that the genetic map obtained from a backcross where the F1 is the female parent is 60% longer than when the F1 is the male parent. It is also clear that environmental factors during meiosis, particularly temperature, can affect the number and distribution of chiasmata, and this could be very critical if the crosses are set up at different times of the year.

Two or more different populations will almost certainly be segregating for different combinations of polymorphic markers, although some at least should be in common. These common ones can be used to overlay the two or more maps and a consensus map produced from an amalgamation of these. Again software is available to produce these consensus maps. Despite the various error-causing factors discussed above a considerable degree of consensus can be produced from such populations and, considerable similarity of chromosome sequence is conserved between even distantly related species, allowing the potential for cross-species genetic transfer. It was stated above that the total map length should be dictated by the chiasma frequency, in so far as this can be accurately measured. When the first few genes were initially put onto genetic maps in the first 40 years of this century, they only covered a small part of the genome – except in wellstudied species such as *Drosophila* and maize. As more and more genes were placed on the map, so the total map lengths increased towards the asymptote dictated by the chiasma frequency. However, during the first decade following the use of molecular markers, the map lengths of some species appeared to exceed that expected, and it was even argued by some that the chiasma theory of recombination might be wrong.

No one would wish to argue that chiasma frequency data are free from error but there would have had to have been quite excessive underscoring to result in the disparity which was occasionally found. It would appear, however, that the excessive lengths were due in part to errors in scoring and to the use of small mapping populations, because subsequent map sizes have shown a progressive decrease towards the predicted asymptote. Scoring errors bias estimates of RF generally upwards, while RFs have to be large to be detectable in small populations. Our own experience with mapping in cereals and brassicas has also shown a progressive decrease in estimated map length as more markers can be found to fill some of the large gaps in the map and as errors are removed from the mapping data.

GENOME ANNOTATION

Genome annotation is the process of attaching biological information to sequences. It consists of two main steps:

1. Identifying elements on the genome, a process called Gene Finding,
2. Attaching biological information to these elements.

Automatic annotation tools try to perform all this by computer analysis, as opposed to manual annotation which involves human expertise. Ideally, these approaches co-exist and complement each other in the same annotation pipeline. The basic level of annotation is using BLAST for finding similarities, and then annotating genomes based on that. However, nowadays more and more additional information is added to the annotation platform. The additional information allows manual annotators to deconvolute discrepancies between genes that are given the same annotation. For example, the seed database uses genome context information, similarity scores, experimental data, and integrations of other resources to provide the most accurate genome annotations through their Subsystems approach. Structural annotation consists in the identification of genomic elements.

- ORFs and their localisation
- Gene structure
- Coding regions
- Location of regulatory motifs

Functional annotation consists in attaching biological information to genomic elements.

- Biochemical function
- Biological function
- Involved regulation and interactions
- Expression

These steps may involve both biological experiments and in silico analysis. A variety of software tools have been developed to permit scientists to view and share genome annotations.

Completion of Genome Projects

When sequencing a genome, there are usually regions that are difficult to sequence (often regions with highly repetitive DNA). Thus, 'completed' genome sequences are rarely ever complete, and terms such as 'working draft' or 'essentially complete' have been used to more accurately describe the status of such genome projects. Even when every base pair of a genome sequence has been determined, there are still likely to be errors present because DNA sequencing is not a completely accurate process. It could also be argued that a complete genome project should include the sequences of mitochondria and (for plants) chloroplasts as these organelles have their own genomes. It is often reported that the goal of sequencing a genome is to obtain information about the complete set of genes in that particular genome sequence. The proportion of a genome that encodes for genes may be very small (particularly in eukaryotes such as humans, where coding DNA may only account for a few percent of the entire sequence).

However, it is not always possible (or desirable) to only sequence the coding regions separately. Also, as scientists understand more about the role of this noncoding DNA (often referred to as junk DNA), it will become more important to have a complete genome sequence as a background to understanding the genetics and biology of any given organism. In many ways genome projects do not confine themselves to only determining a DNA sequence of an organism. Such projects may also include gene prediction to find out where the genes are in a genome, and what those genes do. There may also be related projects to sequence ESTs or mRNAs to help find out where the genes actually are.

Historical and Technological Perspectives

Historically, when sequencing eukaryotic genomes (such as the worm Caenorhabditis elegans) it was common to first map the genome to provide a series of landmarks across the genome. Rather than sequence a chromosome in one go, it would be sequenced piece by piece (with the prior knowledge of approximately where that piece is located on the larger chromosome). Changes in technology and in particular improvements to the processing power of computers, means that genomes can now be 'shotgun sequenced' in one go (there are caveats to this approach though when compared to the traditional approach).

Improvements in DNA sequencing technology has meant that the cost of sequencing a new genome sequence has steadily fallen (in terms of cost per base pair) and newer technology has also meant that genomes can be sequenced far more quickly. When research agencies decide what new genomes to sequence, the emphasis has been on species which have either a relevance to human health (*e.g.* pathogenic bacteria or vectors of disease such as mosquitos) or species which have commercial importance (*e.g.* livestock and crop plants). Secondary emphasis is placed on species whose genomes will help answer important questions in molecular evolution (*e.g.* the common chimpanzee). In the future, it is likely that it will become even cheaper and quicker to sequence a genome. This will allow for complete genome sequences to be determined from many different individuals of the same species. For humans, this will allow us to better understand aspects of human genetic diversity.

MUTATIONS AND EVOLUTION

Every organism has a set of genes and half of the genes of that organism come from each parent. The combinations of the genes causes the variation of individuals within the species. The genes of a butterfly, an ape or a fowl carry the code that determines the appearance and the character of the butterfly, the ape and the fowl.

The genetic code allows an overwhelming variety within the species of the kind. A mutation is basically a gene that has an abnormality in relation to its normal configuration. The abnormality can then be passed to successive offspring, thereby producing a marked difference. There are many different types of mutations. The smallest possible genetic mutation is a 'point-mutation'. This occurs in the DNA when the base-pairs combine with the 'wrong' partner. Multiple point-mutations are common and are found to increase substantially by the effect of mutagens. Mutations are, of course, heritable and these can extend to whole or part chromosomal mutations. Because many genes are affected by a chromosomal mutation, these often have drastic ramifications on the offspring.

MUTATION

In biology, mutations are changes to the base pair sequence of genetic material (either DNA or RNA). Mutations can be caused by copying errors in the genetic material during cell division and by exposure to ultraviolet or ionizing radiation, chemical mutagens, or viruses, or can occur deliberately under cellular control during processes such as meiosis or hypermutation. In multicellular organisms, mutations can be subdivided into germline mutations, which can be passed on to descendants, and somatic mutations. The somatic mutations cannot be transmitted to descendants in animals. Plants sometimes can transmit somatic mutations to their descendants asexually or sexually (in case when flower buds develop in somatically mutated part of plant). Mutations create variations in the gene pool, and the less favorable (or deleterious) mutations are removed from the gene pool by natural selection, while more favorable (beneficial or advantageous) ones tend to accumulate, resulting in evolutionary change.

For example, a butterfly may develop offspring with a new mutation caused say by ultraviolet light from the sun. In most cases, this mutation is not good, since obviously there was no 'purpose' for such change at the molecular level. However, sometimes a mutation may change, say, the butterfly's colour, making it harder for predators to see it; this is an advantage and the chances of this butterfly surviving and producing its own offspring are a little better, and over time the number of butterflies with this mutation may form a large percentage of the species.

Neutral mutations are defined as mutations whose effects do not influence the fitness of either the species or the individuals who make up the species. These can accumulate over time due to genetic drift. The overwhelming majority of mutations have no significant effect, since DNA repair is able to mend most changes before they become permanent mutations, and many organisms have mechanisms for eliminating otherwise permanently mutated somatic cells.

Classification

By Effect on Structure

The sequence of a gene can be altered in a number of ways. Gene mutations have varying effects on health depending on where they occur and whether they alter the function of essential proteins. Structurally, mutations can be classified as:

1. Small-scale mutations, such as affecting a small gene is one or a few nucleotides, including:
   a. Point mutations, often caused by chemicals or malfunction of DNA replication, exchange a single nucleotide for another. Most common is the transition that exchanges a purine for a purine (A "! G) or a pyrimidine for a pyrimidine, (C "! T). A transition

can be caused by nitrous acid, base mispairing, or mutagenic base analogs such as 5-bromo-2-deoxyuridine (BrdU). Less common is a transversion, which exchanges a purine for a pyrimidine or a pyrimidine for a purine (C/T "! A/G). A point mutation can be reversed by another point mutation, in which the nucleotide is changed back to its original state (true reversion) or by second-site reversion (a complementary mutation elsewhere that results in regained gene functionality). These changes are classified as transitions or transversions. An example of a transversion is adenine (A) being converted into a cytosine (C). There are also many other examples that can be found. Point mutations that occur within the protein coding region of a gene may be classified into three kinds, depending upon what the erroneous codon codes for:

i. *Silent mutations*: which code for the same amino acid.
ii. *Missense mutations*: which code for a different amino acid.
iii. *Nonsense mutations*: which code for a stop and can truncate the protein.

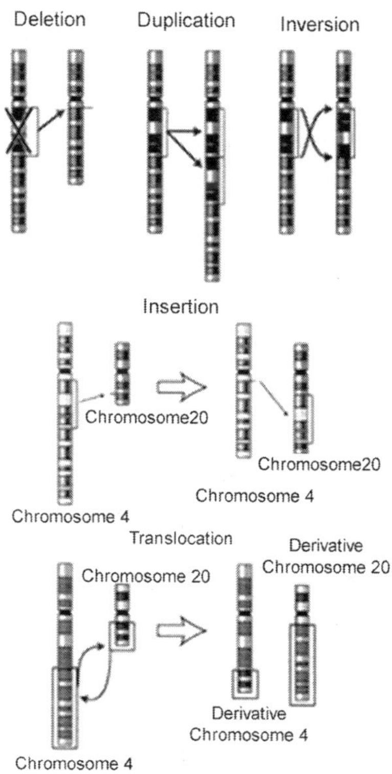

Fig. Five Types of Chromosomal Mutations.

b. Insertions add one or more extra nucleotides into the DNA. They are usually caused by transposable elements, or errors during replication of repeating elements (*e.g.* AT repeats). Insertions in the coding region of a gene may alter splicing of the mRNA (splice site mutation), or cause a shift in the reading frame (frameshift), both of which can significantly alter the gene product. Insertions can be reverted by excision of the transposable element.
c. Deletions remove one or more nucleotides from the DNA. Like insertions, these mutations can alter the reading frame of the gene. They are irreversible.

2. Large-scale mutations in chromosomal structure, including:
   a. Amplifications (or gene duplications) leading to multiple copies of all chromosomal regions, increasing the dosage of the genes located within them.
   b. Deletions of large chromosomal regions, leading to loss of the genes within those regions.
   c. Mutations whose effect is to juxtapose previously separate pieces of DNA, potentially bringing together separate genes to form functionally distinct fusion genes (*e.g.* bcr-abl). These include:
      i. Chromosomal translocations: interchange of genetic parts from nonhomologous chromosomes.
      ii. Interstitial deletions: an intra-chromosomal deletion that removes a segment of DNA from a single chromosome, thereby apposing previously distant genes. For example, cells isolated from a human astrocytoma, a type of brain tumor, were found to have a chromosomal deletion removing sequences between the "fused in glioblastoma" (fig) gene and the receptor tyrosine kinase "ros", producing a fusion protein (FIG-ROS). The abnormal FIG-ROS fusion protein has constitutively active kinase activity that causes oncogenic transformation (a transformation from normal cells to cancer cells).
      iii. Chromosomal inversions: reversing the orientation of a chromosomal segment.
   d. Loss of heterozygosity: loss of one allele, either by a deletion or recombination event, in an organism that previously had two different alleles.

By Effect on Function

- Loss-of-function mutations are the result of gene product having less or no function. When the allele has a complete loss of function (null allele) it is often called an amorphic mutation. Phenotypes associated with such mutations are most often recessive. Exceptions are when the organism is haploid, or when the reduced dosage of a normal gene product is not enough for a normal phenotype (this is called haploinsufficiency).

- Gain-of-function mutations change the gene product such that it gains a new and abnormal function. These mutations usually have dominant phenotypes. Often called a neomorphic mutation.
- Dominant negative mutations (also called antimorphic mutations) have an altered gene product that acts antagonistically to the wild-type allele. These mutations usually result in an altered molecular function (often inactive) and are characterised by a dominant or semi-dominant phenotype. In humans, Marfan syndrome is an example of a dominant negative mutation occurring in an autosomal dominant disease. In this condition, the defective glycoprotein product of the fibrillin gene (FBN1) antagonizes the product of the normal allele.
- Lethal mutations are mutations that lead to a phenotype incapable of effective reproduction.

By Aspect of Phenotype Affected

- Morphological mutations usually affect the outward appearance of an individual. Mutations can change the height of a plant or change it from smooth to rough seeds.
- Biochemical mutations result in lesions stopping the enzymatic pathway. Often, morphological mutants are the direct result of a mutation due to the enzymatic pathway.

Special Classes

- Conditional mutation is a mutation that has wild-type (or less severe) phenotype under certain "permissive" environmental conditions and a mutant phenotype under certain "restrictive" conditions. For example, a temperature-sensitive mutation can cause cell death at high temperature (restrictive condition), but might have no deleterious consequences at a lower temperature (permissive condition).

PLANT GENETIC RESOURCES

Effective conservation and use of plant genetic resources involve asking many questions about the extent and distribution of genetic variation. Only when the appropriate markers and technologies for describing this variation are accessible can such questions be adequately addressed. The most appropriate markers for a given question should: (i) be heritable; (ii) discriminate between the individuals, populations, or taxa being examined; (iii) be easy to measure and evaluate; and (iv) provide results that can be compared with results of similar studies. Molecular markers and assays that meet these criteria were introduced in the 1950s and have proliferated ever since. Use of protein isozymes as genetic markers increased rapidly after their introduction. Nucleic acid markers gained popularity in the 1970s, with the advent of DNA sequencing and restriction fragment analysis. The number of different molecular markers and assays has increased exponentially since the late 1980s, when introduction of the polymerase chain reaction (PCR) made DNA amplification and sequencing possible and affordable.

As the diversity of markers and assays increases, the cost of using them continues to decrease. At the same time, the range of taxa and tissues that can be analysed has expanded. Improved DNA extraction, amplification, and sequencing protocols allow us to analyse ancient DNA, from 13,000-year-old specimens of *Mylodon* (ground sloths) and 17–20 million-year-old *Taxodium* (bald cypress) specimens, and samples as small as single cells or fractions of a seed.

Thus the chances are good that at least one marker assay will be appropriate for any question about variation in any taxa. Furthermore, results of one study can generally be related to results from many other studies. Such opportunities are inevitably accompanied by challenges. None of the challenges are unique to molecular studies, but all become more critical as the number of available marker assays increases. Faced with a bewildering variety of molecular technologies (and their acronyms), choosing a marker and assay that are appropriate for a scientist's particular question, taxa of interest, and practical constraints can be overwhelming.

The temptation is great to use techniques that are recommended by coworkers, are frequently used, for which the most information is available, or are the newest – without adequately evaluating the techniques to determine which is most suitable. Because molecular markers can be measured in virtually all plant taxa and minimal tissue is required, extra care must be taken to verify that the entries in a test array are correctly identified. Since molecular markers are widely applicable and their constraints are not always understood, thoughtful planning is particularly important when designing molecular studies, to ensure that they have an appropriate range of entries, an appropriate number of entries per taxon, and conclusions that stay within the scope of a specific question and test array. In addition, the vast amount of published research using molecular markers complicates the task of thoughtfully relating one study to others with similar marker assays, research questions, and/or taxa.

Another challenge is formulating research objectives that ask critical questions about plant genetic resources. Descriptive studies that use molecular marker assays to examine variation between and within specific taxa can generate valuable information, but some plant conservation questions are best addressed by experimental studies. The wide variety of molecular marker assays available provides new ways to address such questions – if the markers are appropriately used. Such challenges can be met successfully and thoughtful decisions can be made if logical frameworks are available to categorize and compare these valuable molecular tools.

A solid foundation for such comparisons has been provided recently. This chapter builds on that foundation by reviewing both early and recently developed molecular markers and assays, citing examples of their use. The types of questions asked about plant genetic resources are described, as are guidelines for selecting appropriate markers and assays to address these questions. The literature cited here is far from comprehensive, but includes applications of each marker assay in plant genetic research. While the list of molecular techniques will soon be dated, hopefully the framework presented will be helpful for categorizing and comparing new tools as they are made available. The chapter concludes with a perspective on new marker assays as they apply to, and may change, the study of plant genetic variation.

MARKERS AND MARKER ASSAYS

All studies of genetic variation have four primary components: (i) the research question; (ii) the marker used to address it; (iii) the technique used to generate and measure the marker; and (iv) the method used to analyse the data produced. This chapter focuses on the first three components.

PLANT GENETIC RESOURCE

Most questions asked about *ex situ* and *in situ* plant genetic variation fit into one of three categories. While boundaries between them are not always clear, these categories are useful when deciding how to address specific questions. Questions of identity are forensic in nature and ask whether two or more samples are genetically the same.

These questions concern whether genebank accessions are present in duplicate, whether populations are genetically distinct or are geographical subdivisions of one population, and whether genetic change has occurred in an accession or population over time. In contrast, questions of location and diagnostics concern the presence and location of a particular allele or nucleotide sequence in a taxon, genebank accession, *in situ* population, individual, chromosome, or cloned DNA segment.

These questions are asked to locate populations or individuals that have desirable traits; determine whether traits have been lost from a population, or have been transferred from one population or plant to another; and establish the position of markers on physical or genetic maps. The differences between diagnostic and forensic questions are significant, but are not always clarified. Both types of question require qualitative information about specific loci. Yet while diagnostic studies often use single-locus markers, each with few alternative states, forensic questions are commonly addressed with information from many multiallelic loci.

Finally, relationship and structure questions examine relatedness or similarity between genotypes, the amount of genetic variation present, and how variation is distributed between individuals, populations, and taxa. Such questions are used to pinpoint gaps in genebank collections, decide which genotypes are high priority for *in situ* conservation, design germplasm sampling strategies and regeneration programmes, construct core collections, study gene flow, and determine optimum sizes for conserved populations or genebank accessions. These issues are generally addressed with information from many loci. Quantitative data can be used, but qualitative data are more appropriate for some phylogenetic and parentage questions.

Molecular Markers

The appropriate markers for a study can discriminate between entries in an array, but are not so polymorphic that important variation is masked by random noise. Molecular markers range from highly conserved to hypervariable, and can be either proteins or nucleic acids. The nucleic acids used as markers include entire genomes, single chromosomes, fragments of DNA or RNA, and single nucleotides.

A wide variety of nucleic acid fragments are utilized as markers. While some occur once in a genome, others are repeated. Many repeated sequences used as markers are noncoding; others are elements of multigene families. Some repeated sequences are interspersed throughout the genome, either distributed randomly or in clusters. These interspersed repeats are common in plant and animal nuclear genomes, and are found in plant (but not animal) mitochondrial genomes. The chloroplast genome contains a large inverted repeat (IR); most angiosperm chloroplasts have two copies, separated by a short single- copy region. Repeat length and (rarely) loss of one copy can vary between taxa.

Much research at present is focused on repeated sequences that occur in tandem. The classes of tandem repeats are distinguished by the length of the core repeat unit, the number of repeat units per locus, and the abundance and distribution of loci. The names for these classes are themselves varied and have been inconsistently used, but Harding *et al.* (1992) and Tautz (1993) have clarified the nomenclature.

Tandem repeats were first reported in the literature as 'satellites' of DNA, detected in CsCl density gradients as fractions with different GC content than the rest of the genome. These satellites have repeat units that are usually several hundred nucleotides long, with thousands of copies at each of several loci in the nuclear genome. These loci are usually in heterochromatin, often near centromeres. Satellite DNA is present in numerous species. For many satellites, the number of loci and number of repeat units per locus vary between species and higher taxa. Minisatellites (often called variable number of tandem repeat loci, or VNTR loci) are widely used as markers, especially in forensics. The repeat units are usually less than 100 nucleotides long, with tens to hundreds of copies per locus.

Thousands of loci in a genome may have similar core repeat units. The number of repeat units at a minisatellite locus can vary greatly between individuals and populations. First described in humans, minisatellites are found in numerous animal species, often near telomeres. They are also common in plants, and are often associated with satellites and centromeres. The number of repeat units per locus is less variable in plants than in animals, but is still high; plant minisatellites are useful markers for variation between and within species.

MOLECULAR MARKER ASSAYS

Molecular marker assays are generally classified by whether the molecules evaluated are proteins or nucleic acids, and whether the character analysed in a nucleic acid marker assay is the entire genome, a chromosome, a fragment, or a nucleotide. Alternatively, marker assays can be categorized by the type of character measured. Some methods measure quantitative differences between entries in an array. Others measure qualitative characters, each with two or more possible states. Marker assays also differ in the number of loci evaluated per analysis, whether multiple loci are evaluated simultaneously or sequentially, and the type and amount of information needed about the marker loci before conducting the assay.

Protein Marker Assays

When proteins are used as genetic markers, the assumption is made that any variation between proteins reflects heritable variation in their amino acid sequences. This assumption has long been debated, since protein phenotypes can be affected by factors such as post-translational modification, plant phenology, and environmental conditions during plant growth. For many tasks, however, protein marker assays continue to be useful molecular tools. Some assays measure immunological properties of proteins. When antigenic proteins of one plant or animal species are injected into an animal (usually a rabbit), antibodies are produced. Genetic differences between entries in an array can be assessed by comparing how each entry's proteins bind to antibody serum (antiserum) from a reference species.

This binding affinity is measured by the amount of antibody–antigen precipitate produced or by microcomplement fixation (MCF). Each MCF solution contains antiserum from a reference sample, antigen from a test entry, and a protein complement. The complement can only bind to antibody that is bound to the antigen. The amount of unfixed complement is measured; this is used to calculate the amount of unbound antigen, which indicates the extent of variation (the immunological distance) between the entry and the reference sample. After MCF assays are conducted for each entry in an array, these immunological distances are compared between entries. MCF has long been used in vertebrate phylogenetic analysis.

Until recently, immunological markers were seldom used to study plant variation, perhaps because few suitable proteins had been identified. During the past few years, however, monoclonal antibodies to seed proteins of several cereal species have been produced.

Reactivity of these antibodies with antigenic proteins can be evaluated by enzyme-linked immunosorbent assays (ELISAs). Monoclonal antibodies have been used to identify cDNA library clones and describe variation between crop species and cultivars. Other marker assays measure the rate of protein migration through a starch, polyacrylamide, or cellulose acetate gel in response to an electrical current. The migration rate is related to protein size, shape, isoelectric point and/or ionic charge.

In these assays, protein extracts from entries in an array are loaded in adjacent gel lanes and electrophoresed. The gel is then treated with a histochemical stain that reacts with a specific marker protein. Variation between the entries is measured by the positions of their stained proteins on the gel. Many of the proteins used as markers are isozymes – functionally similar forms of an enzyme, encoded by different loci or different alleles at a locus. In general all alleles at a locus are expressed, so allelic isozymes can be analysed as codominant variants.

Isozymes are frequently used to describe variation between and within plant populations, species, and genera; for mapping and diagnostics; and in some cases for identification. When comparing higher taxonomic levels, the probability of genetically different electromorphs migrating to the same gel position is increased.

For closely related taxa and populations, however, isozyme analysis is informative and continues to be refined. Immunoelectrophoretic assays measure both protein migration rate in gels and the antibody specificity of proteins. Antigenic proteins are electrophoretically separated, then are detected by their reactions with antibodies. This technique is valuable for diagnostics, mapping, and describing variation between plant cultivars, species, and higher taxa.

GENETIC ALGORITHMS

Genetic programming is a branch of genetic algorithms. The main difference between genetic programming and genetic algorithms is the representation of the solution. Genetic programming creates computer programmes in the lisp or scheme computer languages as the solution. Genetic algorithms create a string of numbers that represent the solution.

Genetic programming, uses four steps to solve problems:

1. Generate an initial population of random compositions of the functions and terminals of the problem (computer programmes).
2. Execute each programme in the population and assign it a fitness value according to how well it solves the problem.
3. Create a new population of computer programmes.
   - Copy the best existing programmes
   - Create new computer programmes by mutation.
   - Create new computer programmes by crossover(sexual reproduction).
4. The best computer programme that appeared in any generation, the best-so-far solution, is designated as the result of genetic programming.

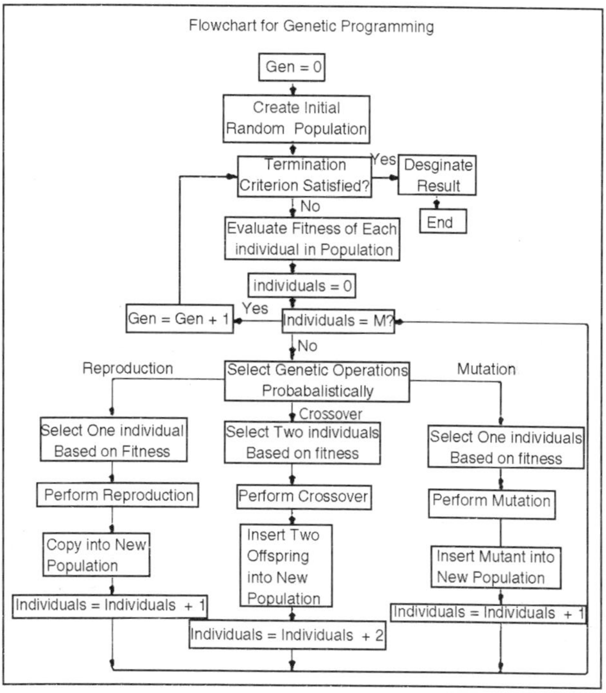

Fig. Genetic Programming Flowchart.

FITNESS FUNCTION

The most difficult and most important concept of genetic programming is the fitness function. The fitness function determines how well a programme is able to solve the problem. It varies greatly from one type of programme to the next.

For example, if one were to create a genetic programme to set the time of a clock, the fitness function would simply be the amount of time that the clock is wrong. Unfortunately, few problems have such an easy fitness function; most cases require a slight modification of the problem in order to find the fitness.

GUN FIRING PROGRAMME

A more complicated example consists of training a genetic programme to fire a gun to hit a moving target. The fitness function is the distance that the bullet is off from the target. The programme has to learn to take into account a number of variables, such as wind velocity, type of gun used, distance to the target, height of the target, velocity and acceleration of the target. This problem represents the type of problem for which genetic programmes are best. It is a simple fitness function with a large number of variables.

WATER SPRINKLER SYSTEM

Consider a programme to control the flow of water through a system of water sprinklers. The fitness function is the correct amount of water evenly distributed over the surface. Unfortunately, there is no one variable encompassing this measurement. Thus, the problem must be modified to find a numerical fitness. One possible solution is placing water-collecting measuring devices at certain intervals on the surface.

The fitness could then be the standard deviation in water level from all the measuring devices. Another possible fitness measure could be the difference between the lowest measured water level and the ideal amount of water; however, this number would not account in any way the water marks at other measuring devices, which may not be at the ideal mark.

MAZE SOLVING PROGRAMME

If one were to create a programme to find the solution to a maze, first, the programme would have to be trained with several known mazes. The ideal solution from the start to finish of the maze would be described by a path of dots. The fitness in this case would be the number of dots the programme is able to find. In order to prevent the programme from wandering around the maze too long, a time limit is implemented along with the fitness function.

FUNCTIONS AND TERMINALS

The terminal and function sets are also important components of genetic programming. The terminal and function sets are the alphabet of the programmes to be made. The terminal set consists of the variables and constants of the programmes. In the maze example, the terminal set would contain three commands: forward, right and left.

The function set consists of the functions of the programme. In the maze example the function set would contain: If "dot" then do x else do y. In the gun firing programme is the terminal set would be composed of the different variables of the problem.

Some of these variables could be the velocities and accelerations of the gun, the bullet and target. The functions are several mathematical functions, such as addition, subtraction, division, multiplication and other more complex functions.

CROSSOVER OPERATION

Two primary operations exist for modifying structures in genetic programming. The most important one is the crossover operation. In the crossover operation, two solutions are sexually combined to form two new solutions or offspring. The parents are chosen from the population by a function of the fitness of the solutions. Three methods exist for selecting the solutions for the crossover operation. The first method uses probability based on the fitness of the solution. If f $(s_i(t))$ is the fitness of the solution Si and,

$$\sum_{j=1}^{M} f\left(s_j(t)\right)$$

is the total sum of all the members of the population, then the probability that the solution Si will be copied to the next generation:

$$\frac{f\left(s_i(t)\right)}{\sum_{j=1}^{M} f\left(s_j(t)\right)}$$

Another method for selecting the solution to be copied is tournament selection. Typically the genetic programme chooses two solutions random. The solution with the higher fitness will win. This method simulates biological mating patterns in which, two members of the same sex compete to mate with a third one of a different sex. Finally, the third method is done by rank. In rank selection, selection is based on the rank, (not the numerical value) of the fitness values of the solutions of the population. The creation of the offsprings from the crossover operation is accomplish by deleting the crossover fragment of the first parent and then inserting the crossover fragment of the second parent.

In genetic programming identical parents can yield different offspring, while in genetic algorithms identical parents would yield identicaloffspring. The bold selections indicate the subtrees to be swapped. An important improvement that genetic programming displays over genetic algorithms is its ability to create two new solutions from the same solution. In the Figure the same parent is used twice to create two new children.

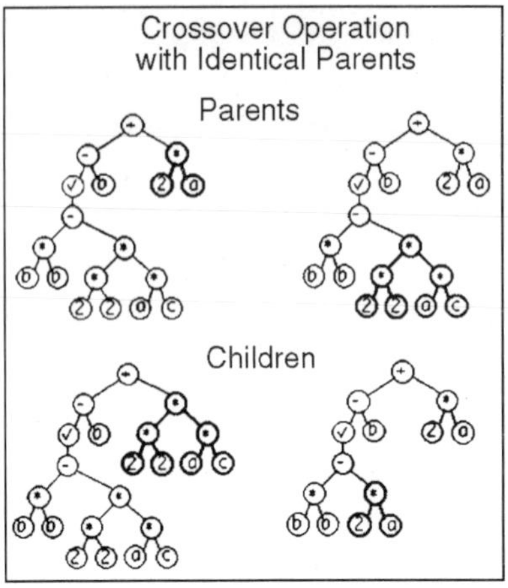

Fig. One of the Main Advantages of Genetic Programming Over Genetic Algorithms.

MUTATION

Mutation is another important feature of genetic programming. Two types of mutations are possible. In the first kind a function can only replace a function or a terminal can only replace a terminal. In the second kind an entire subtree can replace another subtree. Figure explains the concept of mutation:

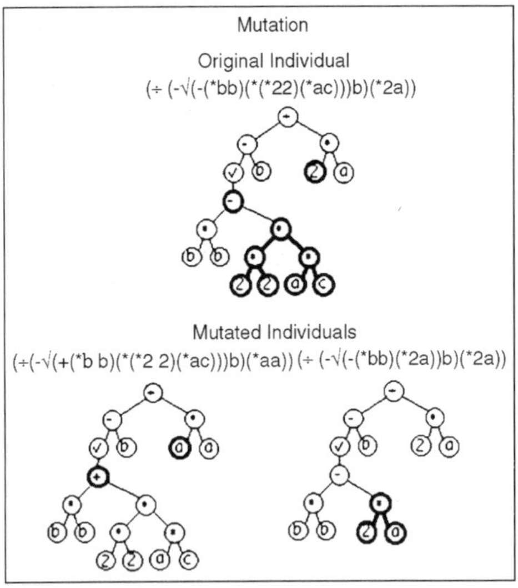

Fig. Two Different Types of Mutations.

The top parse tree is the original agent. The mutation of a single terminal for another single terminal (a). It also illustrates a mutation of a single function (–) for another single function (+). The parse tree on the bottom right illustrates a the replacement of a subtree by another subtree.

Genetic programming is much more powerful than genetic algorithms. The output of the s is a quantity, while the output of the genetic programming is a another computer programme. In essence, this is the beginning of computer programmes that programme themselves. Genetic programming works best for several types of problems. The first type is where there is no ideal solution, (for example, a programme that drives a car). There is no one solution to driving a car. Some solutions drive safely at the expense of time, while others drive fast at a high safety risk. Therefore, driving a car consists of making compromises of speed versus safety, as well as many other variables. In this case genetic programming will find a solution that attempts to compromise and be the most efficient solution from a large list of variables. Furthermore, genetic programming is useful in finding solutions where the variables are constantly changing. In the previous car example, the programme will find one solution for a smooth concrete highway, while it will find a totally different solution for a rough unpaved road.

GENETIC ALGORITHMS - NATURE'S WAY

In the vast eons of time the processes of evolution have created a common basis for all life on Earth - that of DNA or its simpler variant RNA. Both systems share a method of coding with that greatest of human inventions, the computer, yet the mode of operation of the two processes could hardly be more different.

Approaches to Achievement

In our use of computers we first decide an overall goal, we then break down the problem into manageable chunks and write step by step procedures (algorithms) by which to accomplish them. This mirrors our normal approach to life, in which we decide what we wish to do and plan out a way of achieving our ambition, taking into account the resources and time needed. If the plan doesn't work we modify it and try again - a serial, one at a time approach to achievement.

Nature has a different way, brute force and ignorance. If there are a hundred possible options available, well try them all - the first to get there wins. All options are tried simultaneously - a parallel approach to achievement. But where is there ? It just doesn't matter ! At first sight, to human eyes, this is a strange concept - how can you get anywhere useful without knowing where it is ?

TEACHING VERSUS EXPLORING

"Ignorance is bliss" goes the saying and we humans certainly start from a state of ignorance. Each of us can only plan for what we already know, the constraints of our limited knowledge restrict what we can imagine - we often cannot see beyond our assumptions. To overcome these limitations we must rely on being taught new ideas by someone whom we trust, or by stumbling onto new discoveries by accident - in both ways we expand our mental horizons and thus see new possibilities that were previously unthinkable. By adopting nature's method and trying every possible option we can arrive at every destination simultaneously. Thus we can enjoy every possible experience. Wow, what a lifestyle ! Is there a snag ? Of course. We consume a hundred times as many resources for every hundred possible options - and the options increase geometrically (like sand grains doubling for every square on a chessboard), the universe just isn't big enough. Note however, that the conformist social equivalent of everybody doing exactly the same thing (all making the same mistake...), uses just as many resources, yet achieves nothing better - despite the considerableduplication waste involved, our planet isn't big enough for this either !

EVALUATING OPTIONS

But nature has the solution already, tree pruning. Imagine we have two options, each of those has two more and so on, drawing those out we get a picture like a branching tree.

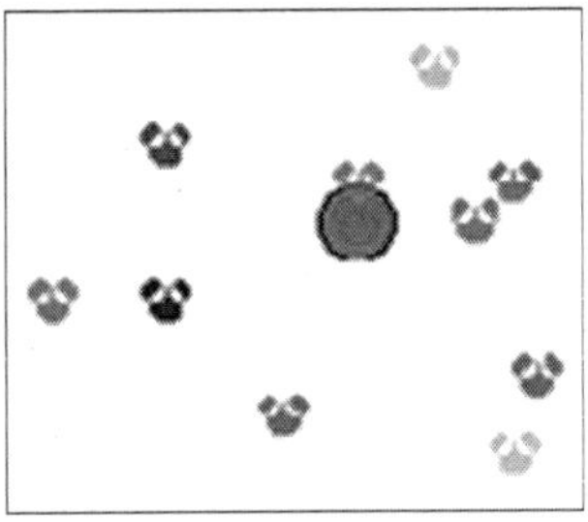

Suppose we impose a test on each option, if the test fails then that option (branch) is deleted (in this illustration we eliminate any mouse genes that fail to find the central food). Once we do this we have pruned the tree of all failed branches and those left passed all our tests (this is similar to how Chess computer programmes work). But which test does nature use ? A very simple one - survival itself. Any organisms unable to survive and reproduce will die out, their genes pruned from the available options.

Can computers operate as nature does ? Yes, this is the general field known as Evolutionary Computation, of which the Genetic Algorithm is the basic technique and can be further expanded to remove the need for an underlying programme.

THE GENETIC ALGORITHM

How does it work? Well, we create a population of individuals, each represented by a chromosome (a collection of genes or characteristics) appropriate to the problem we are trying to investigate. Genes are usually implemented by a series of computer bits, enough to cover all available alternative values. The value of each gene is initially assigned randomly, within the available parameters.

The population is then evaluated to determine how well each individual does the required task. The best members of the population (typically the top 10per cent) then become the parents for the next generation. The genes of the offspring are created by selecting two parents at random and combining part of the chromosome of each, so parent gene combinations ABC mated with DEF may become two offspring ABF and DEC. This mimics the crossover or recombination of sexual reproduction in nature and is repeated for further pairs until a full second population is created. Random changes to individual bits are then made mimicking the natural role of mutation, at some desired rate (say 1 in 100 bits changed).

The resultant population is then evaluated as before, and the process is repeated as many times as desired until the required performance level has been achieved (or no further improvement seems possible). This technique can rapidly covers the space of all possible options and converge on a solution that is beyond the ability of all but the best human programmers, in areas where no conventional solution techniques exist.

COEVOLUTIONARY PROGRAMMING

In the Genetic Programming extension, the genes are the valid statements within a computer language and the chromosome in this case is itself the programme. Additional operations are allowed to join or split programme segments, allowing variable length programmes to evolve. By GA techniques (excluding mutation) we can thus allow a programme to create itself to solve a particular problem, discarding all the failed programmes. This has proved a very powerful way of generating solutions to very difficult problems, but with the drawback that the resultant code is all but incomprehensible to humans ! There is a difficulty here. If our test is too specific, then the programme might be perfect for that one task, but fail completely for very similar tasks. How can we overcome this ? Again by copying nature. Suppose we wish to create a programme to sort a series of numbers. If we have a fixed test sequence of numbers we encounter the problem mentioned, so let us have many sets of numbers, as many as we have candidate programmes.

Programmes can then encounter different tests at each stage and only those capable of a general solution will survive. But we can do better still, let us evolve the tests also by the same process, the test survivors will the ones that defeat a programme. Using this coevolution strategy, the tests get harder as the programmes get better and this bootstrapping feedback technique gives a considerable improvement in speed and efficiency. If rather than programmes and tests we have interacting agents we enter the realm of Artificial Life.

Evolutionary Thinking

Can we use these techniques in human thought processes ? Perhaps we already do. Whenever we think of a task do we not automatically generate many possible solutions and then evaluate them (subconsciously in most cases) against our goal, discarding all those that fail our criteria ? We then refine the survivors (combining the best parts of each) and so on until we have an optimum solution. Nature has again beaten us to it...

Socially a similar process is common. Brainstorming is a popular management technique for generating diverse pseudo-random solutions to a problem, equivalent to nature's mix and match juggling of genes perhaps. The group then evaluates (tests) the options and prunes those failing some criteria, the winners then propagate to the next stage and perhaps ultimately into production. Along with top down (analysis) and bottom up (synthesis) strategies - both goal directed, we also operate a parallel evolution strategy, companies (alternative chromosomes) compete to grow, ideas (memes - cultural genes) compete for attention. Evolution is embedded in our behaviour.

ART AND INTERACTIONS

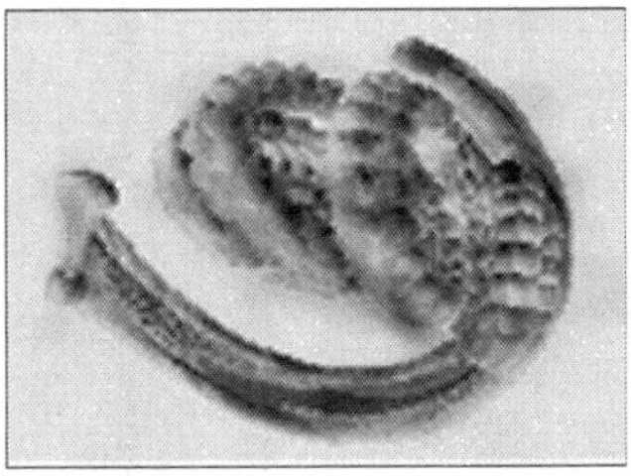

The technique of Genetic Algorithms is widely applicable, even in art. If we code genes to refer to some feature of an image or music we can evolve extraordinary organically based artforms or music. Any system of building blocks can be readily adapted to these procedures and in this way transcend automatically the imagination of its creator. By constraining the system by the desired criteria (tests) we can allow design processes to evolve as a background task - resulting in quite unexpected innovations.

Nature's way. We must however be clear that we are not fully imitating nature here. Our artificial genes (called G-Types) have well defined correspondences with function or form (called P-Types). We also have only one chromosome of very limited length in our genome, whereas natural genotypes contain many chromosomes, each with thousands of genes.In nature genes translate to body form (phenotype or morphology) via a very complex interactive process, the form is itself an emergent phenomena dependent on the interaction of genes with themselves, with the cell chemistry and with the surrounding environment. We cannot predict this process adequately without understanding the nature of emergence. Genetic engineering, whilst highly useful, is a very dangerous process. The side effects of any changes to a such a complex and highly non-linear process as life are, to say the least, unpredictable.

GENETIC PROGRAMMING OPERATORS

The four general genetic operators are: crossover, reproduction, mutation, and inversion. The following overview describes each in brief.

- *Crossover*: Koza considers crossover, along with reproduction, to be the two foremost genetic operations. It is mainly responsible for the genetic diversity in the population of programmes. Similar to its performance under genetic algorithms, crossover operates on two programmes (a binary operator), and produces two child programmes. Two random nodes are selected from within each programme and then the resultant "sub-trees" are swapped, generating two new programmes. These new programmes become part of the next generation of programmes to be evaluated.

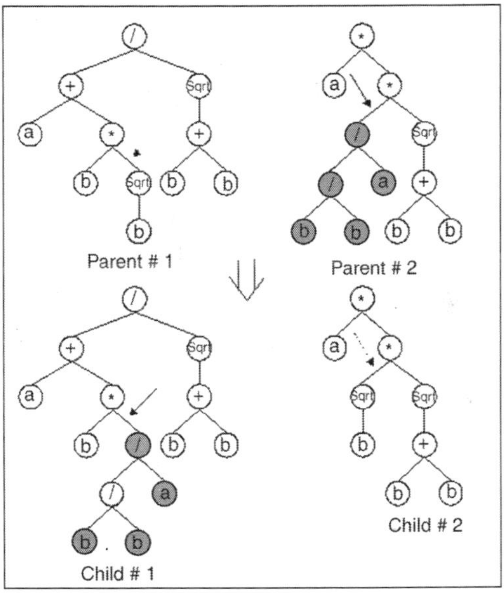

Fig. The Effect of a Crossover Operation.

- *Reproduction*: Reproduction, the second main operation, is performed by simply copying a selected member from the current generation to the next generation. An interesting note about convergence of the population using the two main operators is made by Koza ... in genetic programming, when an individual incestuously mates with itself (or copies of itself), the two resulting offspring will, in general, be different.... As before, the Darwinian reproduction operation creates a tendency towards convergence; however, in genetic programming, the crossover operation exerts a counterbalancing pressure away from convergence. Thus, convergence of the population is unlikely in genetic programming.

- *Mutation*: With genetic algorithms, mutation becomes an important operator which provides diversity to the population. As Koza points out, however, mutation is relatively unimportant in the new environment, because the dynamic sizes and shapes of the individuals in the population already provide diversity, and as stated above, the population should not converge. Thus, mutation can be considered as a variation on the crossover operation.

Index

I

L

M

N

P

R

S

T

U

V

W